Microgrids

Microgrids

Special Issue Editors

Mònica Aragüés-Peñalba
Andreas Sumper

MDPI • Basel • Beijing • Wuhan • Barcelona • Belgrade

MDPI

Special Issue Editors
Mònica Aragüés-Peñalba
Universitat Politècnica de
Catalunya
Spain

Andreas Sumper
Universitat Politècnica de
Catalunya
Spain

Editorial Office
MDPI
St. Alban-Anlage 66
4052 Basel, Switzerland

This is a reprint of articles from the Special Issue published online in the open access journal *Applied Sciences* (ISSN 2076-3417) from 2018 to 2019 (available at: https://www.mdpi.com/journal/applsci/special_issues/Micro_grids).

For citation purposes, cite each article independently as indicated on the article page online and as indicated below:

LastName, A.A.; LastName, B.B.; LastName, C.C. Article Title. *Journal Name* **Year**, *Article Number*, Page Range.

ISBN 978-3-03921-868-4 (Pbk)
ISBN 978-3-03921-869-1 (PDF)

Contents

About the Special Issue Editors

Mònica Aragüés-Peñalba received her degree in Industrial Engineering (major in Electricity) from the Escola Tècnica Superior d'Enginyeria Industrial de Barcelona (ETSEIB) of the Technical University of Catalonia (UPC) Barcelona, Spain, in 2011. In 2016, she obtained her Ph.D. in Electrical Engineering from the UPC. Since 2010, she has been part of CITCEA (Center for Technological Research in Static Converters and Drives), within the Department of Electrical Engineering at the UPC, collaborating in industrial and research projects related to the integration of renewable energy sources, especially wind power and solar photovoltaic power. She has been Lecturer at the Electrical Engineering Department at UPC since 2018 and is also an IEEE Member. She has collaborated in projects focused on the operation and control of transmission systems (HVDC and HVAC) for grid integration of offshore wind, operation and control of photovoltaic power plants, optimal microgrid operation, and grid integration of renewables in distribution grids.

Andreas Sumper was born in Villach, Austria. He received his Dipl.-Ing. degree in Electrical Engineering from the Graz University of Technology in Austria in 2000, and his Ph.D. degree from the Universitat Politecnica de Catalunya, Barcelona, Spain, in 2008. From 2001 to 2002, he was Project Manager for innovation projects in the private sector. In 2002, he joined the Center for Technological Innovation in Static Converters and Drives (CITCEA) at the Universitat Politecnica de Catalunya. At the Department of Electrical Engineering, he was Assistant Professor from 2006 to 2009, Lecturer from 2009 to 2014, and Associate Professor from 2014 to 2018. He is now Full Professor at the Escola Superior d'Ingenieria Industrial de Barcelona (ESTEIB), Universitat Politècnica de Catalunya. His research interests are renewable energy, digitalization of the power grid, micro- and smartgrids, power system studies, and energy management.

applied sciences

MDPI

Editorial

Special Issue on Microgrids

Mònica Aragüés Peñalba * and Andreas Sumper

CITCEA-UPC, Department of Electrical Engineering, Universitat Politecnica de Catalunya, 08028 Barcelona, Spain; andreas.sumper@upc.edu

* Correspondence: monica.aragues@upc.edu

Received: 29 October 2019; Accepted: 31 October 2019; Published: 5 November 2019

1. Introduction

Integration of renewable energy sources in the electrical power system is key for enabling the decarbonization of that system. The connection of renewable generation to the electrical system is being performed in a centralized form (large renewable power plants like wind or solar power plants connected at the transmission system) and in a decentralized manner (through the connection of dispersed generation connected at the distribution system). The connection of renewable generation at distribution levels, together with other generating sources as well as energy storage systems (the so-called DER, Distributed Energy Resources) close to consumption sites, is promoting the development of microgrids: DER installations that have the capability to operate grid connected and grid isolated. The uncertainty and variability of the renewable energy sources that integrate microgrids, as well as the need for coordination with other energy sources, pose challenges in the operation, protection, control, and planning of microgrids. The five selected papers published in this Special Issue propose solutions to address these challenges.

2. Conclusions

The authors from [1] propose overload control strategies for four-wire inverters in low voltage AC (Alternating Current) microgrids. The developed strategies provide a fast and appropriate fault current limitation in both operation modes, grid connected and grid isolated. The strategies are validated through simulations using Matlab/Simulink and real experimental results are obtained from CENER (The National Renewable Energy Centre) experimental ATENEA four-wire AC microgrid, showing time responses in the order of two-three grid cycles for all cases.

The authors from [2] propose a reliability evaluation method for multi-energy microgrids, understood as energy systems with multiple energy vectors that can operate autonomously. A reliability factor is integrated in a planning economic model for these types of systems. The impacts of several equipment configuration schemes on planning and reliability are addressed. A planning-operation optimization model is proposed to ensure the energy supply, determining the output power of the generating and storage units of the microgrid.

In [3], the optimal operation of isolated microgrids, taking into account frequency constraints, is addressed. In particular, a new stochastic optimization method is designed to maximize photovoltaic generation in microgrids combining photovoltaic generation, diesel generation, and energy storage. The optimization problem is formulated including a minimum frequency constraint, which is obtained from a dynamic study considering maximum load and photovoltaic power variations. To maintain the mixed integer linear formulation of the optimization problem, this constraint is defined through a linear regression. Three complete days are simulated to verify the proper behavior of the system under the proposed optimization scheme. The system is validated in a laboratory-scaled microgrid.

While [1–3] focus on a single microgrid, [4] proposes a hierarchical optimization method for the energy scheduling of multiple microgrids connected to the distribution grid with participation in the energy market. The optimization procedure is separated into two stages. The first stage is focused on the optimal operation of each microgrid in the next hour and uses a mixed integer programming formulation. The second stage uses the output from the first and allows the market operator to establish an internal price incentive mechanism (based on Stackelberg Game theory) for the next hour. The goal of the energy market operator is to maximize its profits, taking into account the demand responses of the microgrids. It is shown that based on this optimization, the microgrid operator and the energy market operator can achieve larger benefits.

Last, but not least, [5] focuses on the load frequency control of islanded microgrids consisting of diesel engines, renewable sources, and storage devices. For developing the proposed control, the concept of fractional calculus is combined with sliding mode control. Hardware-in-the-loop tests show that the proposed controller allows frequency fluctuations to be avoided and ensures a more robust operation of the microgrid compared to other techniques.

Acknowledgments: The Editors would like to thank all those that contributed to the publication of this special issue. This issue would not have been possible without the scientific contributions of the authors, time dedication and accuracy of reviewers and excellent editorial team of Applied Sciences.

References

1. Heredero-Peris, D.; Chillón-Antón, C.; Pagès-Giménez, M.; Montesinos-Miracle, D.; Santamaría, M.; Rivas, D.; Aguado, M. An Enhancing Fault Current Limitation Hybrid Droop/V-f Control for Grid-Tied Four-Wire Inverters in AC Microgrids. *Appl. Sci.* **2018**, *8*, 1725. [CrossRef]
2. Ge, S.; Li, J.; Liu, H.; Sun, H.; Wang, Y. Research on Operation–Planning Double-Layer Optimization Design Method for Multi-Energy Microgrid Considering Reliability. *Appl. Sci.* **2018**, *8*, 2062. [CrossRef]
3. Vidal-Clos, J.; Bullich-Massagué, E.; Aragüés-Peñalba, M.; Vinyals-Canal, G.; Chillón-Antón, C.; Prieto-Araujo, E.; Gomis-Bellmunt, O.; Galceran-Arellano, S. Optimal Operation of Isolated Microgrids Considering Frequency Constraints. *Appl. Sci.* **2019**, *9*, 223. [CrossRef]
4. Rui, T.; Li, G.; Wang, Q.; Hu, C.; Shen, W.; Xu, B. Hierarchical Optimization Method for Energy Scheduling of Multiple Microgrids. *Appl. Sci.* **2019**, *9*, 624. [CrossRef]
5. Esfahani, Z.; Roohi, M.; Gheisarnejad, M.; Dragičević, T.; Khooban, M. Optimal Non-Integer Sliding Mode Control for Frequency Regulation in Stand-Alone Modern Power Grids. *Appl. Sci.* **2019**, *9*, 3411. [CrossRef]

applied sciences

MDPI

Article

An Enhancing Fault Current Limitation Hybrid Droop/V-f Control for Grid-Tied Four-Wire Inverters in AC Microgrids

Daniel Heredero-Peris [1,*,†], Cristian Chillón-Antón [1,†], Marc Pagès-Giménez [1,†], Daniel Montesinos-Miracle [1,†], Mikel Santamaría [2], David Rivas [2] and Mónica Aguado [2]

[1] Centre d'Innovació Tecnològica en Convertidors Estàtics i Accionaments (CITCEA-UPC), Departament d'Enginyeria Elèctrica, Universitat Politècnica de Catalunya, ETS d'Enginyeria Industrial de Barcelona, Avinguda Diagonal, 647, Pl. 2, 08028 Barcelona, Spain; citcea@citcea.upc.edu or cristian.chillon@citcea.upc.edu (C.C.-A.); marc.pages@teknocea.cat (M.P.-G.); montesinos@citcea.upc.edu (D.M.-M.)

[2] Renewable Energies Grid Integration Department, CENER (Renewable Energy National Centre of Spain), 31621 Navarra, Spain; info@cener.com or msantamaria@cener.com (M.S.); drivas@cener.com (D.R.); maguado@cener.com (M.A.)

* Correspondence: daniel.heredero@citcea.upc.edu; Tel.: +34-93-401-6855

† These authors contributed equally to this work.

Received: 30 July 2018; Accepted: 19 September 2018; Published: 22 September 2018

Abstract: Microgrid integration and fault protection in complex network scenarios is a coming challenge to be faced with new strategies and solutions. In this context of increasing complexity, this paper describes two specific overload control strategies for four-wire inverters integrated in low voltage four-wire alternating current (AC) microgrids. The control of grid-tied microgrid inverters has been widely studied in the past and mainly focused on the use of droop control, which hugely constrains the time response during grid-disconnected operation. Taking into account the previous knowledge and experience about this subject, the main contribution of these two proposals regards providing fault current limitation in both operation modes, over-load capability skills in grid-connected operation and sinusoidal short-circuit proof in grid-disconnected operation. In the complex operation scenarios mentioned above, a hybrid combination of AC droop control based on dynamic phasors with varying virtual resistance, and voltage/frequency master voltage control for grid-(dis)connected operation modes are adopted as the mechanism to enhance time response. The two proposals described in the present document are validated by means of simulations using Matlab/Simulink and real experimental results obtained from CENER (The National Renewable Energy Centre) experimental ATENEA four-wire AC microgrid, obtaining time responses in the order of two-three grid cycles for all cases.

Keywords: microgrids; control strategies; three-phase four-wire systems; fault current limitation

1. Introduction

In the coming years, it is expected that classical electrical grids will drive forward to a smarter, more flexible, reliable, efficient and bidirectional format leading to a more complex framework. All of these benefits should be supported by an appropriate infrastructure. In this context, microgrids [1,2], and mainly alternating current (AC) microgrids, play a key role in a new electrical paradigm pushed by the increasing penetration of Distributed Energy Resources (DER). This paradigm will deal with the variability and unpredictability associated with DERs and local demand fluctuations. This versatility generates a way to delay the renovation of an aged infrastructure that cannot withstand an existing rising demand [3]. A new outline, constituted by several interconnected AC/DC (direct current) microgrids or nano-grids [4], conventional energy sources, and loads will create future viable smartgrids [5].

Microgrids can provide potential economic and environmental benefits, but their implementation implies great technical difficulties in control, energy/power management and protection. Some authors focus on the low level loops for the inverters' operation; AC droop [6–9], voltage and current control loops [10–12]. Other ones concentrate on the high management level based on cooperative distributed strategies [13,14] or optimal-smart operation [15–18]. Considering the previously mentioned antecedents in the literature, the authors of this paper consider it very relevant to pay attention to fault protection [19,20] and secure operation in seamless transference between operation modes [12,21–23].

On one hand, in traditional AC four-wire distribution systems, protective device coordination during faults is achieved by selecting appropriate circuit-breaker current–time characteristics under clear regulations. This choice does not imply intercomponent communication [21,24] and assumes high short-circuit power levels [24]. However, the situation is the opposite in the case of microgrids based on power electronics, the over-load capability being, hereinafter Fault Current Limitation (FCL), constrained. In a microgrid context, it should be adaptive and fast in terms of voltage and current limitations [25], and should be able to behave sinusoidally to not affect the response of conventional protective breakers.

On the other hand, inverters in a microgrid can play two main roles: one as a controlled voltage source and the other as a controlled current source [26], and can adopt two control hierarchies, master–slave or peer-to-peer [27]. In addition, in the peer-to-peer hierarchy, the conventional AC droop voltage control strategy, based on the steady-state or quasi-static power transference model between AC sources [6,28], is a widely applied alternative to face both operation modes (grid-(dis)connected) and parallellize various inverters, but generates poor dynamics, mainly in grid-disconnected operation. This last situation get worse with the typical use of low-pass filters to emulate synchronous generator mechanical inertias [28]. The combination of droop-based control with the use of virtual impedances is a widespread mechanism to support the soft-start challenge under a peer-to-peer hierarchy [29,30]. Furthermore, conventional AC droop control presents low adaptability when the operation point differs significantly from the planned rated point. In this situation, the recent use of dynamic phasors can be used to improve the adaptability of conventional AC droop control strategies [31].

The motivation of this paper regards considering strategies that provide FCL capabilities in both operation modes. It can be found in [24,32,33] examples based on separating the fault from the grid rapidly but without an AC fault management strategy. However, previous examples as well as others, focus on faults at the DC-link of the inverter [34], observing the effect on the AC side. On the other side, others references concentrate directly on DC microgrids [35–37]. The last two scenarios mentioned above are far away from this paper target. Another studied solution is to face FCL under the assumption that the inverter is droop-based in both operation modes. In [38,39], examples for three-phase three-wire microgrids are exposed. Different types of short circuits are studied in [38], but the fault current limitations offer dynamics of about 200 ms to achieve steady-state, while in [39] only a tretrapolar short-circuit is evaluated obtaining time responses of about 100 ms. In [40], some results are just simulated for AC droop-based four-wire systems. Finally, other alternatives exist applied to series filters or obtaining FCL by changing the output inductances as detailed in [41,42].

The main contribution of the present paper is the demonstration by simulation and real results of the advantages and flexibility of a fast time response hybrid combination of voltage control techniques that ensures proper FCL capabilities introducing two strategies for this purpose: one for each operation mode. This allows for applying specific control strategies and solutions in a context of increasing complexity scenarios avoiding the use of generic solutions that are not always the most appropriate ones.

Firstly, the AC droop control based on dynamic phasors is adopted for the grid-connected mode, but a master voltage-frequency (V/f) control strategy is embraced for the grid-disconnected mode, offering better dynamic responses during the grid-disconnected operation thanks to disabling the AC droop loop. The use of a variable virtual resistance supports not only soft-starts but also the transference between the operation modes. Varying virtual resistance enhances the voltage restoration during the transference.

Secondly, two control strategies will assist the FCL capability of the inverter connected to an AC microgrid. An over-load supervisor is proposed to characterize and limit the over-load magnitude, providing thermal recovery when extra current has to be dispatched during grid-connected mode. In addition, a short-circuit proof strategy supports the operation of the inverter under different short-circuit situations and fault clearances for grid-disconnected operation, obtaining time responses below 60 ms. To achieve the same response from the protective devices' viewpoint, the short-circuit proof algorithm allows for maintaining current and voltage sinusoidal and totally controlled, as required in a microgrid operation framework.

All strategies are thought to be compatible with a four-wire microgrid because, as detailed in [43–45], this is the proper solution for addressing independently three-phase current control and facing imbalances. These strategies become an efficient way to face common situations in low voltage microgrids interfaced with distribution networks.

Thus, the paper is organized as follows. Section 2 defines the system. Section 3 describes the proposed strategies for the FCL and the adaptive virtual resistance mechanism adopted for a seamless smooth transference. Section 4 presents the simulated results using Matlab/Simulink (R2017b, The MathWorks, Inc., Natick, MA, USA, 1984) and exposes the experimental ones validated in the experimental ATENEA four-wire microgrid at The National Renewable Energy Centre (CENER). Finally, Section 5 provides the conclusions.

2. System Definition

The following sub-sections define the experimental ATENEA microgrid and the converter considered in this paper.

2.1. The Experimental ATENEA Microgrid

The Renewable Energy Grid Integration Department in CENER (National Renewable Energy Centre of Spain) has developed and deployed a microgrid (ATENEA) placed in the Rocaforte industrial area (town of Sangüesa, Navarra, Spain) according to the interest in an industrial test scenario and environment. The generation equipment of the facility can be seen in Figure 1.

It consists of an AC architecture with total installed power of about 120 kW that can supply part of the Wind Turbine Test Laboratory (LEA), electric loads and Rocaforte industrial lighting area. It also can be used as a test-bench for different generation and storage technologies and control strategies. The generation equipment available in the facility can be seen in Figure 1.

The ATENEA microgrid structure is based on an AC low voltage three-phase, four-wire bus (400 V, 50 Hz) connected to all of the equipment. This experimental microgrid has two configurations: grid-connected and grid-disconnected. Its main objective is to manage generation and demand in order to obtain high ratios of energy self-efficiency. In grid-connected mode, the microgrid is connected to the network and the V/f performance is fixed by its own network. In the grid-disconnected scenario, one of the converters linked with an energy storage system (flow battery, Valve-Regulated Lead-Acid (VRLA) battery, Li-ion battery) or a diesel generator is configured to form the grid, mastering the V/f. In this way, the ATENEA microgrid adopts a master–slave control in the grid-isolated configuration and the slaves work under a PQ control, P and Q being active and reactive power, respectively.

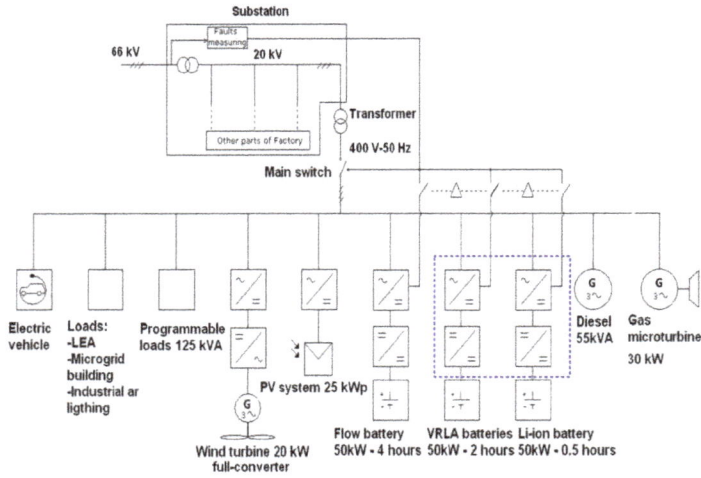

Figure 1. Diagram of the ATENEA microgrid at The National Renewable Energy Centre (CENER). A blue dashed box indicates the two possible direct current–alternating current (DC–AC) converters that can implement the proposed strategies.

2.2. The Converter

According to Section 2.1, a two-stage converter is considered to interface with the VRLA or the Li-ion 50 kW batteries of Figure 1 (blue dashed box) with the four-wire microgrid. It has been decided to use a common two-level three-leg power stack to maintain homogeneity for both power stages (see Figure 2).

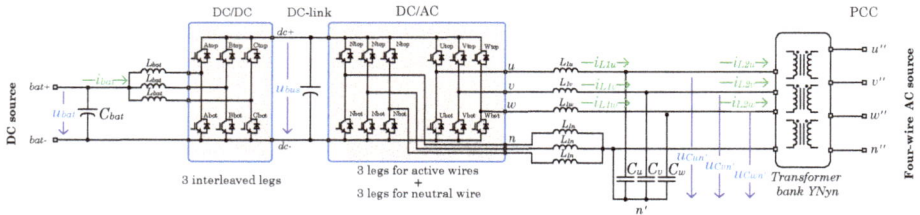

Figure 2. Scheme of the full direct current–direct current (DC–DC) and DC–AC proposed converter.

In consonance with the available storage technologies, a DC voltage u_{bat} range from 150 to 500 VDC is considered. With this configuration, a DC–DC converter interfaces with this wide DC voltage range by means of a three-phase interleaved topology. By the use of an interleaved topology, the output inductive filter size is split and reduced, so it makes for easier operation and maintenance tasks. At the same time, high power converters could be designed with lower current modules, reducing the voltage and current ripple in the DC-link and decreasing the power capacity of the inductors [46]. Thus, as cited in [47], the reliability of an interleaved DC–DC converters increases compared with conventional one leg devices.

In order to manage AC unbalanced loads, an inverter topology able to control any current sequence is required. The inverter stage is constituted by two three-leg bridges: one dedicated for the active phases and the other one for the neutral wire. This configuration allows for controlling each line current independently using optimized modulation techniques such as Space Vector Pulse Width Modulation (SVPWM) [48]. An LCL-type coupling filter completes the inverter where an isolation

6

free-flux YNyn transformer bank is assumed as part of the LCL-type filter providing galvanic isolation and offering different possible neutral schemes [49].

3. Control Strategies

The following sub-sections define the main control strategies of the proposed converter in Section 2 that will be applied in the experimental ATENEA microgrid.

3.1. DC–DC Interleaved Converter Control

DC–DC power stage is not the aim of this paper, but, in the context of this paper, it has been considered relevant to briefly mention its high level control details. In the operation context described in this document, the DC–AC inverter operation requires a proper DC-link voltage level to hold an adequate behaviour under unbalances or nonlinear needs. In addition, it has been considered that the DC energy storage system presents a large voltage range. Due to this voltage wide voltage range, an uncontrolled constant DC-link voltage level can affect inverter's operation. The purpose of the DC–DC converter is to step-up the voltage of the storage system and provide an autonomous way to regulate the DC-link voltage level against different AC requirements.

The DC–DC converter is controlled by means of two nested control loops, as shown in Figure 3a. The inner control loop manages the battery inductor, L_{bat}, and limits the maximum desired battery current, i_{bat}. As detailed in Section 2.2, an interleaved topology is chosen, so a $2\pi/3$ rad/s shifted phase PWM strategy is used with the same duty cycle for all the converter's legs. The outer control loop controls the DC-link voltage, u_{bus}. Because of the unbalanced nature of AC connected loads, low frequency voltage ripples (at $\simeq$100 Hz) in the DC-link can be severe. This voltage control loop needs a high bandwidth to overcome this ripple and keep constant u_{bus}. Thus, an adaptive 100 Hz notch filter [50] is used in the feedback control chain for this purpose to quickly compensate for any variation around the fundamental rated grid frequency (50 Hz).

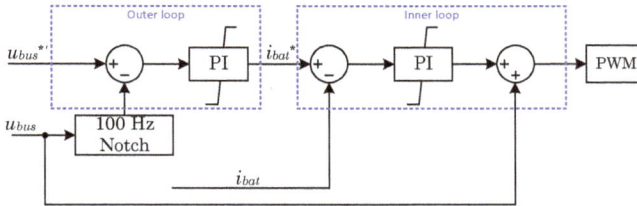

(**a**) DC–DC interleaved converter nested control loops

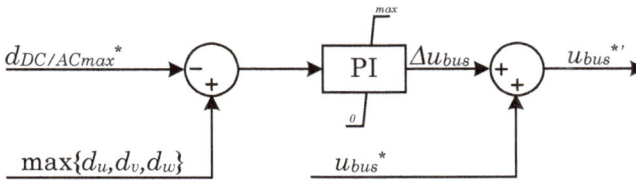

(**b**) Adaptive DC-link voltage reference generator (higher level preventive controller)

Figure 3. Interleaved DC–DC converter control schemes. Proportional Integral (PI).

To avoid over-modulation situations in the DC–AC stage, a preventive controller is suggested at the higher level of the DC–DC control scheme in order to provide an increment of u_{bus}, Δu_{bus}, to the rated DC-link voltage. Thus, the reference $u_{bus}^{*\prime} = u_{bus}^{*} + \Delta u_{bus}$ is generated where u_{bus}^{*} is the desired standard DC-link voltage, as can be deduced from Figure 3b. The maximum available duty cycle of the inverter, $d_{DC--ACmax}^{*}$, is compared with the maximum of the duties of the inverter's active phases (d_u, d_v and d_w). If there is not enough DC bus for the inverter, the DC reference is stepped up to its maximum threshold, taking into account the limitations of the DC-link. In another case, Δu_{bus} is equal to zero in the steady-state as a consequence of the lower saturation limit of the PI controller, as shown in Figure 3b. Thanks to this method, the DC-link voltage is increased as required only under imminent over-modulation situation.

3.2. DC–AC Four-Leg Converter Control

The DC–AC power stage is responsible for operating the AC side in the two operation modes defined in Section 2.1: grid-connected and grid-disconnected. In this paper, and following the operation philosophy of ATENEA's operation, a strategy based on maintaining the voltage source behaviour in both modes has been adopted. However, and as a difference from the solutions proposed in the literature review of Section 1, the grid-connected operation adopts an AC droop control, but, in grid-disconnected operation, the voltage behaviour is maintained disabling the external droop loop. In this last case, the voltage/frequency control is assumed by the inverter being the voltage master of the microgrid. In this scenario, the obtained dynamics are less limited in time-response due to the absence of droop constraints.

As defined in Section 2, the inverter is based on a three-phase four-leg topology to be fully compatible with ATENEA four-wire microgrid. This allows the inverter to be controlled by means of three independent single phase systems in order to provide direct, indirect and homopolar sequence control capability. Each phase has its own master AC droop control (only for grid-connected mode) and two inner cascaded stationary frame controllers [51] for the voltage ($u_{C_{xn'}}$) and current ($i_{L_{1x}}$) loops, x phase being u, v or w. The inner loops are tuned considering [49,52] and the tuning values are presented in Section 4.

3.2.1. Control Assumptions

Classical AC droop control operation principles are obtained from the steady-state equations that describe the power flow between two AC voltage sources connected by an inductive line, as widely detailed in [6,28]. A predominant resistive behaviour is adopted by using the virtual impedance concept [29,30] to provide a reliable relationship between the sets active-reactive power and voltage-frequency, being as independent as possible from the grid impedance. The virtual resistance concept is described in Figure 4. In Figure 4a, u_C designs the controlled AC voltage, u_{PCC} is the voltage of Point of Common Coupling (PCC), R_2 the physical equivalent series resistance of L_2, and R_v is the value of the virtual resistance (see Figure 2).

The adaptive virtual resistance concept is currently used for hot-swapping (soft-start) [6] and to smooth the effect of grid fluctuations. In this case, it is also applied to enhance the transference between operation modes, as is later shown in Section 4.2. It should be noted that the virtual resistance should be disabled progressively to not affect the operation of a pure V/f strategy in the grid-disconnected mode. Figure 4b illustrates the proposed behaviour of the R_v module during and between the operation modes. Furthermore, as it has been aforementioned, this fact makes it possible to improve the time response performance in grid-disconnected mode thanks to the master voltage role change between the mains and the converter.

It is possible to deduce the dynamic AC control droop schemes depicted in Figure 5 under a resistive behaviour assumption and considering dynamic phasors [31]. This control schemes are the basis control schemes adopted for the grid-connected operation in the paper. In Figure 5, G_{ctrl} represents the transfer function of the AC droop control law between the node A that is the AC controlled capacitor and a node B corresponding to the PCC. For the G_{ctrl}, it is assumed that a τ_f time constant for emulating mechanical inertias of synchronous generators [28]. P^*/Q^*-P/Q are the active/reactive powers set-points and measured values, and U_A and U_B represent the voltage at the mentioned nodes A and B. The ω_{UB} is the angular frequency at the PCC, θ_{UA} and θ_{U2} are again the phases at nodes A and B, respectively. Finally, R and L are the total resistive and inductive part between A and B. Note that R is the addition of the real equivalent series resistance of the wiring, the output transformer involved and the forced virtual part R_v.

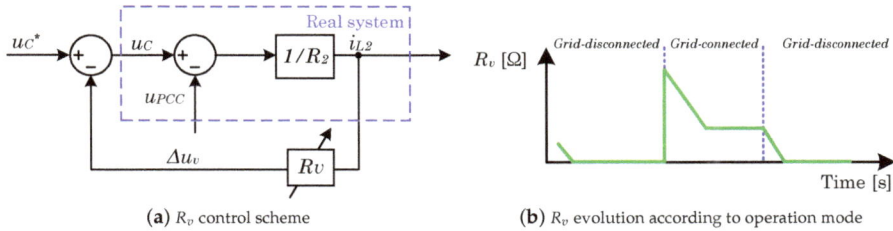

(**a**) R_v control scheme

(**b**) R_v evolution according to operation mode

Figure 4. The virtual resistance R_v operation.

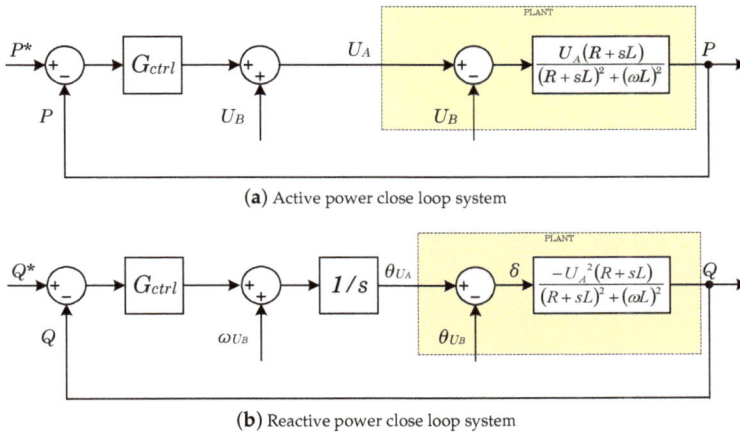

(**a**) Active power close loop system

(**b**) Reactive power close loop system

Figure 5. Resistive case power close loop scheme under dynamic phasors.

3.2.2. Power Over-Load Supervisor for the Grid-Connected Mode

As it is well known, the inertial and over-load capability of power electronics is limited. Moreover, it is common to consider pulsed drain currents of two to three times the continuous drain current for about 1 ms. On the contrary, in traditional AC systems, the rotary machines and transformers can be over-loaded up to 20–30 times in a timescale of minutes [24]. In this case, it has been considered to over-size the converter allowing a certain over-load capability. This fact provides a more flexible interaction between the inverter and the AC microgrid. The oversizing is achieved using high current switching devices. This choice allows for a more compact converter placing the burden of oversizing only on the cost of switching devices, a not really sensitive part today. Thus, the cooling system is designed to suit the nominal power. However, the thermal time lag of the used cooling method is usually enough and within the range of seconds to minutes. Furthermore, the use of thermal masses such as aluminium plates, or the consideration of phase change materials are options to increase the thermal inertia [53,54], being a good trade-off between cost and volume, if required.

In order to manage the over-load capability, the maximum AC current is handled by a power over-load supervisor algorithm based on thermal criteria. This current limitation is achieved by means of the apparent power s (in per unit). An over-load observer, ol_o, limits the power per phase. The over-load observer fulfills this task through the formula

$$ol_o = \int_0^t \left(i^{*2} - 1 \right) dt \tag{1}$$

based on the i^2t computation. In this sense, the term ol_o is an indicator of the over-load energy exchanged, i^* is the desired current and t is the time interval of the over-current. The over-load algorithm is managed according to the state diagram shown in Figure 6. Time t begins when $|s^*| > 1$, s^* being the maximum apparent power per phase reference. When $|s^*| > 1$, the observer enters to the *Wake-up* step, starting to compute ol_o (Equation (1)) and ol_t (accumulative time under the over-load situation). If the ol_o value reaches zero, the observer returns to a *Sleep* step, where ol_o and ol_t are reset. If the ol_o is bigger than zero, depending on the present s value, the accumulated time ol_t is incremented or maintained. Keeping the ol_t time constant, it is ensured that the inverter is not over-loaded intermittently producing possible thermal degradations. In the case that the ol_t term becomes higher than a pre-set threshold $T_{max\,ol}$, the system evolves to the *Prolonged over-load error* step and s is limited to 0.8. In this way, ol_o is forced to decrease until zero. Then, the system evolves to the *Sleep* step again.

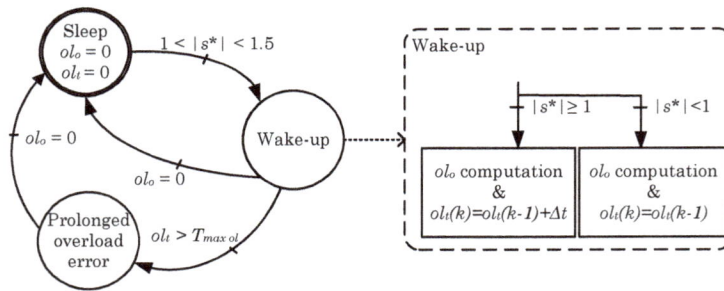

Figure 6. Scheme of the power over-load supervisor algorithm.

3.3. Short-Circuit Proof Algorithm for Grid-Disconnected Mode

As detailed in Section 2, the inverter operates as the voltage master when the ATENEA microgrid is grid-disconnected operated. The inverter must provide sinusoidal currents even under an over-load or short-circuit occurrence reducing the voltage accordingly. In this way, the aim of the microgrid inverter and this paper is to generate a totally controlled short-circuit power regulating sinusoidal currents. Right after, this information can be used to (re)configure certain protection devices' distributed thresholds along the AC microgrid [25]. In this sense, the time response of the breaker that feeds the fault should be affected minimally.

Figure 7 shows the proposed block diagram of the short-circuit proof mechanism. The conventional voltage and current stationary frame controllers can be seen in black, i.e., Proportional Resonant with Harmonic Compensator (PRHC) controllers, where the subscript u and i refer to voltage and current, respectively. u_C^* is the objective AC capacitor voltage, U_C the current AC capacitor voltage, and u_I the inverters' output voltage. i_{L1} and i_{L2} are the inverters' output inductance and grid coupling inductance. The superscript * designs the reference, and the subscript rms is the root mean square (rms) computed value.

The blue parts are added with respect to a conventional voltage-current nested loop for the short-circuit proof enhancement. The algorithm is based on the per phase rms value of the current reference on the AC side, i_{L1}^*, and two factors k_1 and k_2. The first factor k_1 allows for dynamically regulating the voltage set-point to attenuate it under short-circuit or high over-load situations, as follows:

$$k_1 = \begin{cases} 1, & \text{if} & i_{rms} \leq I, \\ (K+1) - \frac{i_{rms}K}{I}, & \text{if} & I < i_{rms} < (K+1)\,I, \\ 0, & \text{if} & i_{rms} \geq (K+1)\,I, \end{cases} \tag{2}$$

where I is the maximum desired rated output current (at the L_1 inductance). The parameter K in k_1 calculation allows for adapting the system response speed to face the over-load or short-circuit occurrence.

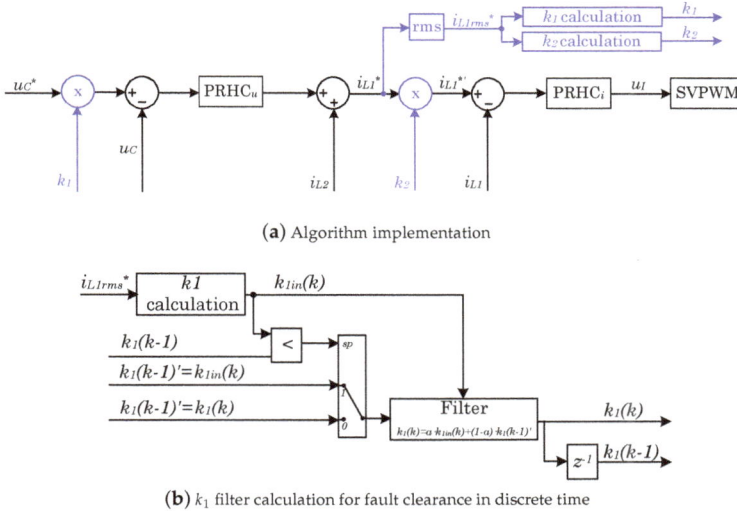

(**a**) Algorithm implementation

(**b**) k_1 filter calculation for fault clearance in discrete time

Figure 7. Control schemes of the sinusoidal wave short-circuit proof algorithm.

The second factor k_2 limits the current to the rated value when the fault situation appears as

$$k_2 = \begin{cases} 1, & \text{if} \quad i_{\text{rms}} < I, \\ \frac{I}{i^*_{L1rms}}, & \text{if} \quad i_{\text{rms}} \geq I. \end{cases} \tag{3}$$

If only the previous algorithm is applied, the behaviour under the fault clearance is undesired. This is due to the fast response of the k_1 factor. To avoid this kind of undesired dynamics, when the current $k_{1in}(k)$ value is higher than the previous computed one $k_1(k-1)$, i.e., this criterion is used as a fault recovery indicator, the applied $k_1(k)$ gain for the inner current control reference is filtered according to Figure 7b.

4. Results

This section describes a 90 kVA converter with a 50% over-load capability. A set of simulations developed in Matlab/Simulink and experimental results obtained at ATENEA microgrid are showed for the validation of the aforementioned control contributions. Hereinafter, the ITI curve [55] defined by the Information Technology Industry Council is considered as a pattern of acceptable time-duration/magnitude voltage transients.

4.1. The Converter Set-Up

The converter presented in Section 2 is based on three Semikron IGD-2-424 power stacks (Semikron, Nuremberg, Germany). The control is implemented into two TMS320F2809 DSP-based control boards, one dedicated to control the general operation state machine and the DC–DC interleaved converter and the other committed to controlling the inverter. Control strategies are executed at 8 kHz. The hardware and software relevant parameters for the DC–AC and the DC–DC converters are summarised in Tables 1 and 2. Figure 8 shows a picture of the full converter. Furthermore, all short-circuit faults and recoveries are generated using a switch-line breaker (ABB OT200) and considering wires with less than 0.1 mΩ.

Table 1. Interleaved DC–DC (direct current–direct current) converter parameters. PI (Proportional Integral).

	Parameter	Value	Units
Adaptive DC-link PI controller	k_p	0.043	
	k_i	1.43	
Adaptive 100 Hz filter [50]	Adaptive coefficient μ	0.05	
	Attenuate B coefficient of cut-off frequency	4	
PI Voltage controller	k_p	3.5	
	k_i	70	
PI Current controller	k_p	0.16	
	k_i	33.75	
DC–DC converter	Switching & control frequency	8	kHz
	L_{bat} (each interleaved inductance)	400	μH
	C_{bat}	420	μF
	$C_{DC-link}$	7.2	mF

Table 2. Four-wire DC–AC (direct current–alternating current) converter parameters. PRHC (Proportional Resonant with Harmonic Compensator).

	Parameter	Value	Units
Droop controller	m for active power loop	0.000003	
	n for reactive power loop	0.000004	
	k_i for reactive power loop	0.0009	s
	t_f Low-pass filter (LPF) constant	0.1	
PRHC Voltage controller	k_p	0.27	
	k_{i_0}	0.26	
	k_{i_3}	0.001	
	k_{i_5}	0.001	
PRHC Current controller	k_p	0.7468	
	k_{i_0}	3.93	
	k_{i_3}	0.1	
	k_{i_5}	0.04	
Fault current limiters	K	0.9	
	I	130	A
	α (over-load filter parameter)	0.01	
Virtual impedances	R_v (initial-state grid-connected)	1.0	Ω
	R_v (steady-state grid-connected)	0.2	Ω
	R_v (steady-state grid-disconnected)	0	Ω
	R_v change ratio	−0.16	Ω/s
DC–AC converter	Switching & control frequency	8	kHz
	L_1 (active phases & neutral wire)	250	μH
	C (star connected)	350	μF
	L_2 (leakage transformer inductance)	70	μH

4.2. Simulated Results

This section focuses on demonstrating by the use of simulations the virtual resistance contribution to a fast and seamless transference between operation modes, the power over-load supervisor operation, and the fault current limitation strategy for an ideal (<1 mΩ impedance) short-circuit occurrence and clearance.

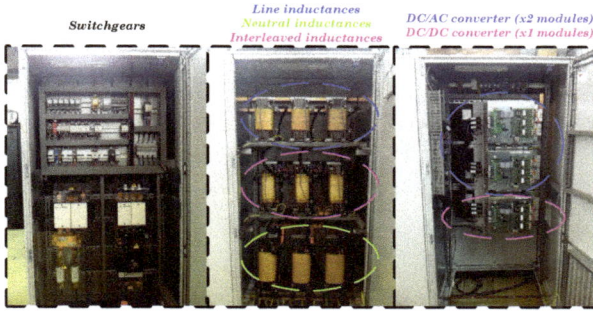

Figure 8. Converter installed in the ATENEA experimental microgrid.

4.2.1. Virtual Resistance Effect on the Transference

For this analysis, a transference is forced to 50 ms time. A main switch to segregate the mains from the microgrid with a turn-on/off delay of 40/120 ms has been assumed.

In Figure 9, the effect of the R_v value on the voltage $u_{u''n''}$ and the delivered current i_{L2} can be observed when a grid-connected to grid-disconnected transference takes place. In the initial situation, P_u^* is set at 30 kW before t = 50 ms and a local load of 1.81 Ω is connected at any time. The use of an initial R_v allows for smoothing the transference in terms of voltage and currents.

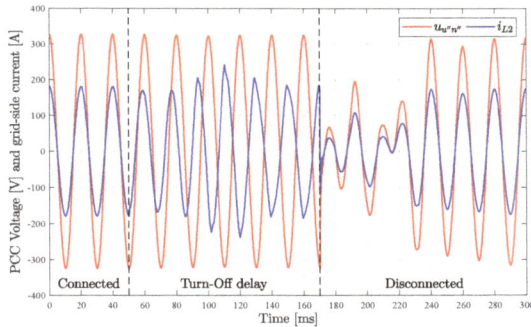

(**a**) R_v is set to 0 Ω immediatly after the grid-disconnection

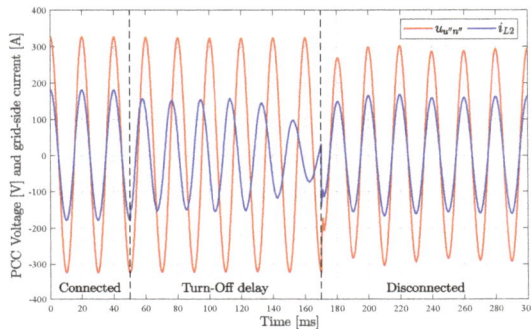

(**b**) R_v is set to 0 Ω progressively at a -0.16 Ω/s ratio after the grid-disconnection

Figure 9. R_v (0.2 Ω initially) effect during grid-connected to grid-disconnected transference at $t = 50$ ms.

In Figure 10, the effect of the R_v value on the voltage $u_{u''n''}$ and the delivered current i_{L2} can be observed when a reconnection transference takes place. In this case, in the initial situation, a load of 1.81 Ω is connected and maintained after the reconnection. No PQ references are considered in this case. As in the disconnection case, the use of an initial R_v smooths the transference in terms of voltage and currents. The use of high R_v values helps to extinguish the i_{L2} current faster but creates a voltage dip. However, this virtual resistance allows for making the system less sensitive due to interfacing with a virtual current limiter during the reconnection process. Then, a trade-off R_v value has to be used when the reconnection occurs. According to the mentioned reasons, an R_v equal to 1 Ω is suggested as a proper compromise value, also used in the experimental validation in Section 4.3.1.

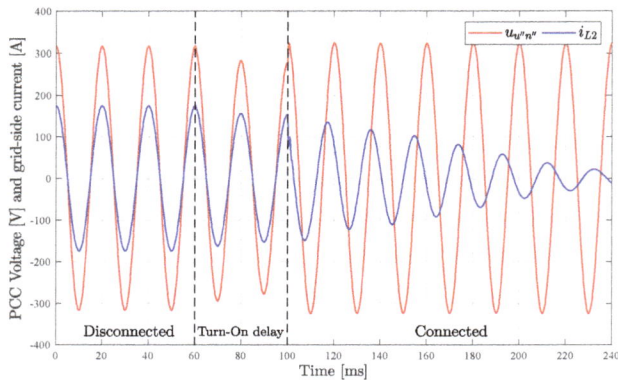

(**a**) R_v is set to 0.2 Ω after the grid-connection

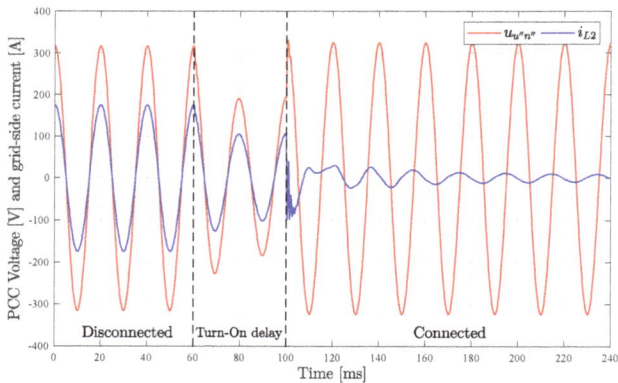

(**b**) R_v is set to 1 Ω after the grid-connection

Figure 10. R_v (0 Ω initially) effect during grid-disconnected to grid-connected transference at t = 50 ms.

4.2.2. Power Over-Load Supervisor

Figure 11 shows the behaviour of the power over-load supervisor strategy presented in Figure 6. In Figure 11, s^* refers to the power set-point received from any external manager, while s_{int}^* represents the inner reference managed by the mentioned over-load supervisor strategy. Note that the current tracking is supposed to operate properly, as it is illustrated in Section 4.

In Figure 11, three different study cases framed in yellow can be observed. The first one refers to a short-time over-load demand. The second one accumulates two over-load queries. Finally, the third one evaluates a prolonged over-load target. The maximum over-load time, $T_{max\,ol}$, is set to 1 s.

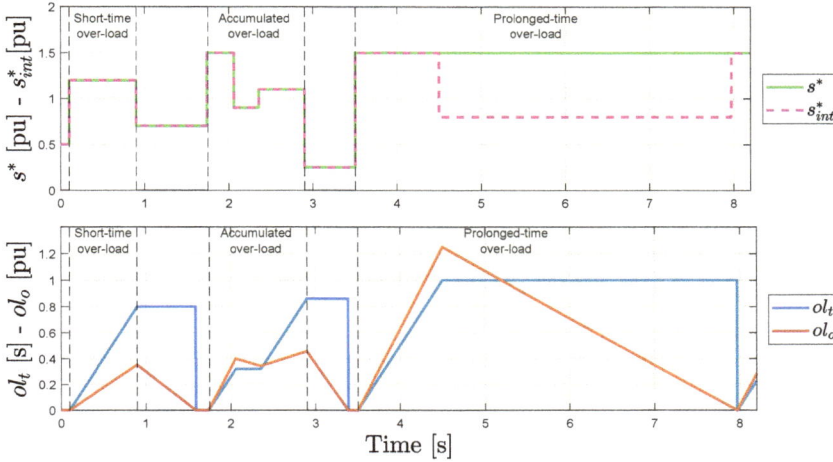

Figure 11. Example of the power over-load supervisor algorithm behaviour.

Regarding the short-time scenario, when s^* is higher than 1, the o_{l_t} and o_{l_o} values increase. When s is lower than 1, the o_{l_t} time is held while o_{l_o} decreases progressively according to Equation (1). In this sense, the generated thermal stress on the cooling system is limited. This situation is exemplified in the accumulated over-load region. Although the set-point s is reduced transiently, the o_{l_t} time continues increasing due to not achieving o_{l_o} equal to zero. Lastly, for a prolonged-time, an over-load region can be observed as to how the over-load strategy limits the inner set-point s_{int}^* to 0.8 [pu]. This situation is produced when o_{l_t} reaches a pre-configured $T_{max\,ol}$ and remains unaltered until the o_{l_o} reaches zero. Note that the maximum over-load time, $T_{max\,ol}$, can be configured for each phase according to any specific design of the cooling system. Thus, the 1 s previously selected is just an example.

4.2.3. Short-Circuit Proof Algorithm

Scenario 1—Phase to neutral short-circuit. Figure 12a,b show the behaviour of the voltage and current of phases u'' and v'' when a unipolar $u''n''$ short-circuit is generated and recovered, respectively. It can be seen that the voltage goes to zero when the fault appears maintaining the current limited with a sinusoidal waveform. When the fault is recovered, the voltage increases progressively without producing any problematic over-voltage. In both cases, the time response is less than 60 ms.

Scenario 2—Phase to phase short-circuit. Figure 13a,b show the behaviour of the voltage and current of phases u'' and v'' when a bipolar $u''v''$ short-circuit is generated and recovered. It can be deduced $i_{L2u} = -i_{L2v}$ and $u_{u''n''} = u_{v''n''}$, as can be also observed in Figure 13b. This case is particularly interesting because, although the current is properly managed in the steady-state, it can be seen that the voltage does not go to zero after the fault.

When the fault occurs, it is possible that, in one of the two involved phases, its voltage control action, $PRHC_u$ output, plus the short-circuit current, i_{L2}, adds up to more than in the other case (see Figure 7b). The phase with more errors rapidly produces a k_1 gain that moves from one to zero. As the other phase operates with higher k_1 values, it starts to control the current without necessarily a k_1 gain equal to zero. This means without the correspondent phase-to-neutral voltage equal to zero.

When the fault is recovered, the voltage increases progressively without producing any problematic over-voltage. As in the unipolar case, the transients are resolved in less than 2–3 grid cycles.

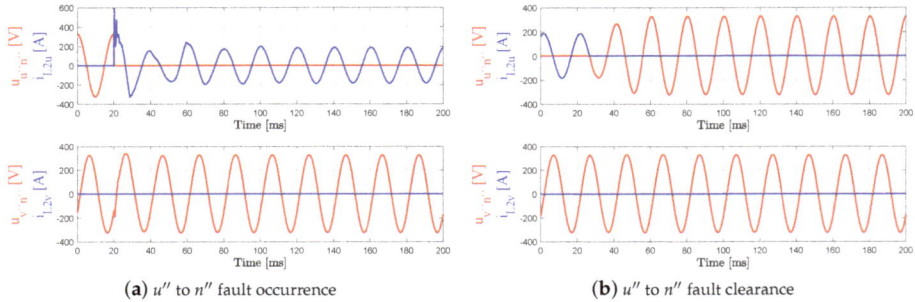

(**a**) u'' to n'' fault occurrence (**b**) u'' to n'' fault clearance

Figure 12. Scenario 1 (simulated phase to neutral short-circuit fault).

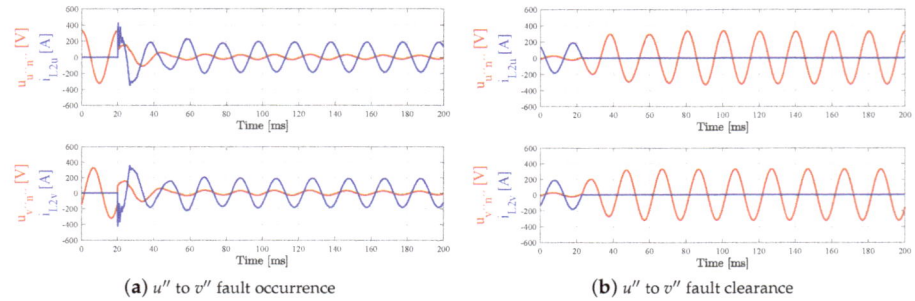

(**a**) u'' to v'' fault occurrence (**b**) u'' to v'' fault clearance

Figure 13. Scenario 2 (phase to phase short-circuit fault).

Scenarios 3–4—Three-phase short-circuit and Three-phase to neutral short-circuit. Figures 14 and 15 show the behaviour of the voltage and current of phases u'' and v'' when a tripolar between phases or a tetrapolar short-circuit is enforced and recovered, respectively. As in the previously studied cases, the behaviour of the current and voltage follows similar time responses.

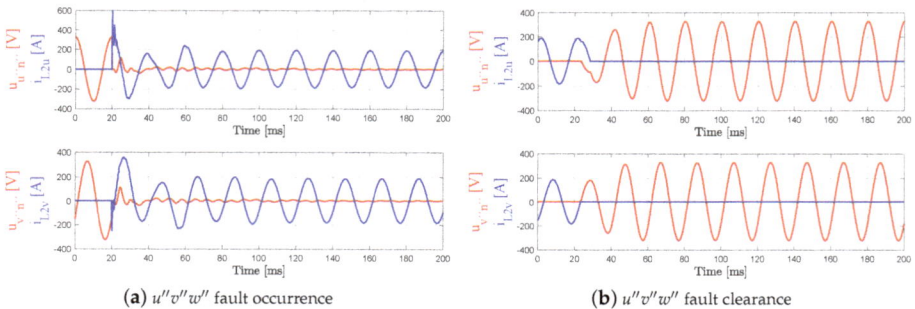

(**a**) $u''v''w''$ fault occurrence (**b**) $u''v''w''$ fault clearance

Figure 14. Scenario 3 (three-phase short-circuit).

(a) $u''v''w''$ to n'' fault occurrence

(b) $u''v''w''$ to n'' fault clearance

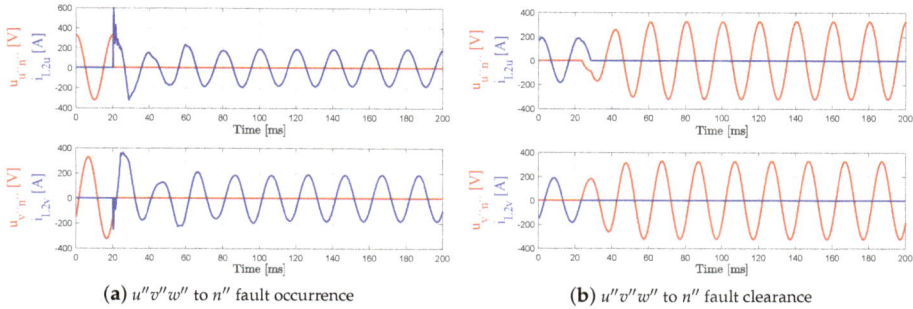

Figure 15. Scenario 4 (three-phase to neutral short-circuit).

4.3. Experimental Results

This section is divided into three subsections to validate the suggested proposals along the paper and check the simulation results presented in the previous section.

4.3.1. Virtual Resistance Effect on the Transference

In Section 4.2.1, the theoretical effectiveness of the virtual resistance algorithm to smooth the transference between grid-(dis)connected modes was demonstrated by simulations. For the experimental validation, only the R_v value of 1 Ω has been used when the reconnection occurs, as detailed in Section 4.2.1. Figure 16 shows the experimental results for two scenarios, grid-connected to grid-disconnected and vice versa. A load of 1.81 Ω has been considered.

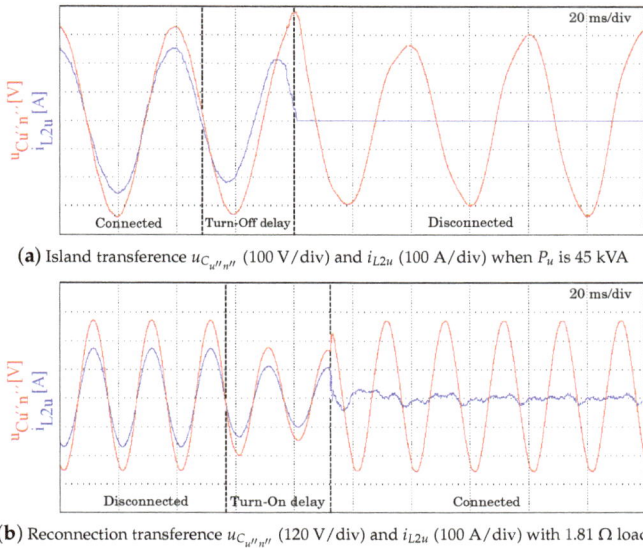

(a) Island transference $u_{C_{u''n''}}$ (100 V/div) and i_{L2u} (100 A/div) when P_u is 45 kVA

(b) Reconnection transference $u_{C_{u''n''}}$ (120 V/div) and i_{L2u} (100 A/div) with 1.81 Ω load

Figure 16. Scenario 1 (**a**) and 2 (**b**). Microgrid transitions between operation modes.

Scenario 1—Transition from grid-connected to grid-disconnected at maximum power without local loads. Phase u is delibering 45 kW while the other two phases are with null PQ requests. Then, a non-intentional disconnection to start operating in grid-disconnected is done. Figure 16a shows the proper behaviour of the voltage, creating a short-duration dip that meets the ITI curve. After the transition, the delivered current goes to zero and the voltage of the microgrid is maintained.

Scenario 2—Transition from grid-disconnected to grid-connected with load. A reconnection procedure is intentionally done with a high current local load. Figure 16b allows for observing that, after the reconnection, the $u_{C_{u''n''}}$ voltage suffers a dip during two cycles, but the current i_{L2} is extinguished just after the reconnection is finished. The experimental results are close to the simulation shown in Figure 10b, validating the exposed method.

In both cases, the PCC voltage suffers an alteration, the transition from grid-disconnected to the grid-connected mode being more critical. However, in both cases, the voltage alterations are resolved by control in less than 40 ms, meeting the ITI curve requirements. This trade-off permits a safe system connection even in weak grids while complying the regulation requirements.

4.3.2. Four Quadrant Control Capability

The objective of this section is showing the capability of the inverter to control unbalanced phase currents as mentioned in Section 1 and required for the proper use of the over-load supervisor strategy suggested in Section 4.2.2. In this scenario, the inverter is operating in grid-connected mode with rated (non over-loaded) unbalanced *PQ* set-points. The set-points per phase are $P_u = P_w = 30$ kW, $P_v = -30$ kW and Q_{uvw} all nulls. Figure 17 shows that the inverter is able to synthesize non-balanced currents from non-balanced *PQ* references.

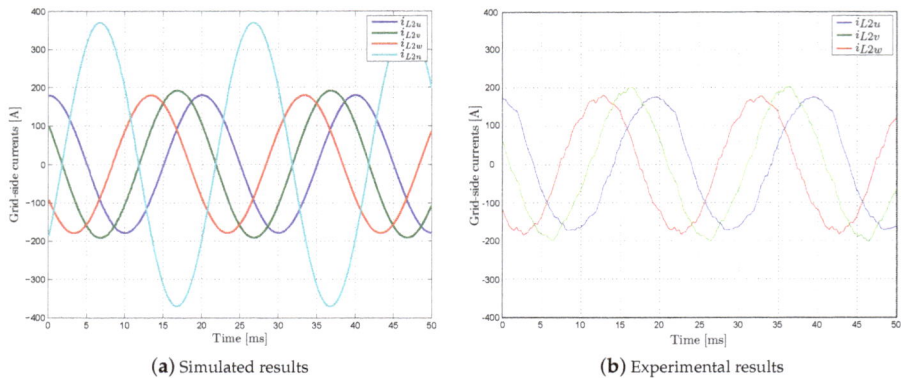

(a) Simulated results   **(b)** Experimental results

Figure 17. Inverter in grid-connected mode. Unbalanced set-point: $P_u = P_w = 30$ kW, $P_v = -30$ kW and $Q_{uvw} = 0$ kvar.

Figure 17a shows a simulation of the expected active phase and neutral wire output currents, respectively. Figure 17b presents the captured oscilloscope active currents. It is demonstrated that there is no problem to track unbalanced *PQ* set-points. Thus, it is possible to ensure that if an over-load supervisor manages the inner converter references, called s_{int}^* in Figure 11, the inverter can provide autonomous over-load capability per phase.

4.3.3. Short-Circuit Proof Algorithm

In this section, the four scenarios simulated in Section 4.2.3 are reproduced in the experimental platform to validate the real capabilities of the proposed fault current limitation strategy. In the following lines, it is demonstrated that the time response for the fault current limitation action and voltage recovery offers superior time responses than in the current literature [38–40].

Scenario 1—Phase to neutral short-circuit. Figure 18 shows the behaviour of the voltage and current of phases u'' and v'' when unipolar $u''n''$ is produced. Analogously with Figure 12a, it can be seen in Figure 18a that, when the fault appears, the control maintains the current limited with a sinusoidal waveform presenting a minor oscilation that is resolved in less than 30 ms. Figure 12b shows the system behaviour when the fault is cleared, recovering the nominal voltage value progresively in less than two grid cicles, in the same way as the simulation reproduced in Figure 12b.

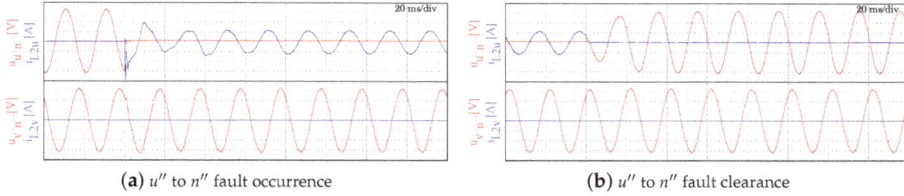

(**a**) u'' to n'' fault occurrence   (**b**) u'' to n'' fault clearance

Figure 18. Scenario 1 (phase to neutral short-circuit fault). $u_{x''n''}$ (100 V/div) and i_{L2x} (150 A/div).

Scenario 2—Phase to phase short-circuit. Figure 19 shows the behaviour of the voltage and current of phases u'' and v'' when a bipolar $u''v''$ short-circuit is generated and recovered, analogous to simulations shown in Figure 13. In Figure 19a, it can be seen that the initial transient of current u'' and v'' is less than two times the maximum current, achieving steady-state values in less than 30 ms. When the fault is recovered, the voltage increases progressively achieving the steady-state in less than 40 ms as shown in Figure 19b.

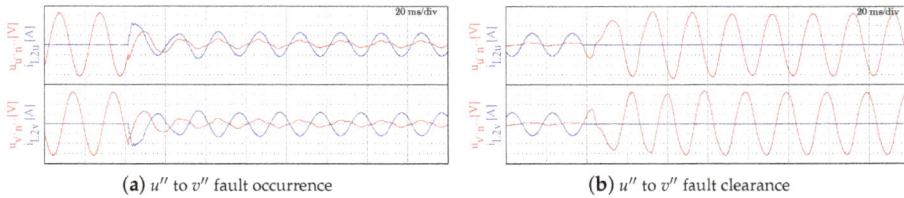

(**a**) u'' to v'' fault occurrence   (**b**) u'' to v'' fault clearance

Figure 19. Scenario 2 (phase to phase short-circuit fault). $u_{x''n''}$ (100 V/div) and i_{L2x} (150 A/div).

Scenarios 3–4—Three-phase short-circuit and Three-phase to neutral short-circuit. Figures 20 and 21 show the behaviour of the voltage and current of phases u'' and v'' when a tripolar and tretrapolar fault is produced. As in the simulated cases shown in Figures 14 and 15, the control algorithm limits the fault current in the expected time, being less than 40 ms. The fault recovery is also achieved rapidly, in less than two grid cycles.

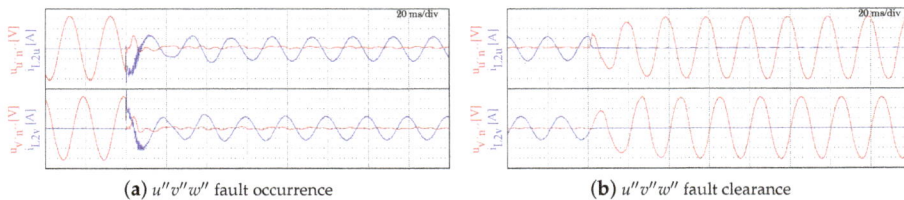

(**a**) $u''v''w''$ fault occurrence   (**b**) $u''v''w''$ fault clearance

Figure 20. Scenario 3 (three-phase short-circuit). $u_{x''n''}$ (100 V/div) and i_{L2x} (150 A/div).

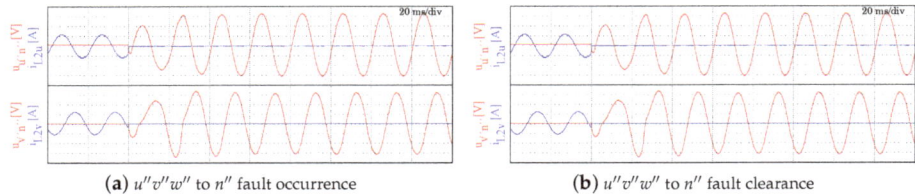

(**a**) $u''v''w''$ to n'' fault occurrence   (**b**) $u''v''w''$ to n'' fault clearance

Figure 21. Scenario 4 (three-phase to neutral short-circuit). $u_{x''n''}$ (100 V/div) and i_{L2x} (150 A/div).

5. Conclusions

This paper has presented two control strategies to achieve a fast and proper fault current limitation for four wire microgrid inverters. Both strategies are supported by a dynamic virtual resistance mechanism that allows for a fast transference between grid-(dis)connected operation modes.

Concerning the AC side inverter operation, a seamless transference between grid-(dis)connected modes has been ensured obtaining short transients, below three grid cycles. For this purpose, a combination of an AC droop control based on dynamic phasors when grid-connected and master voltage/frequency control when grid-disconnected has been proposed. It has been demonstrated not only that this is a valid option to maintain the voltage behaviour between operation modes, but also it is able to make the grid-disconnected operation independent from the AC droop control loop constraints. Thus, the transients in grid-disconnected mode are only limited by the settling time of the voltage control loop. Furthermore, a varying virtual resistance is suggested for achieving the mentioned seamless transference, disabling the resistance value progressively when entering in grid-disconnected operation. It has been illustrated that the proper selection of the steady-state virtual resistance and the used variation ratio helps to avoid voltage sags in the transient phase between the operation modes, allows for being less sensitive to the reconnection process and, finally, provides a predominant resistive behaviour for the grid-connected operation.

An over-load manager supervisor strategy based on thermal criteria has been introduced. This over-load strategy provides an increase in managing flexibility to a microgrid inverter in a grid-connected mode. The over-size challenge is then delegated to higher current switching devices, a not really cost sensitive component today in power converters. Thus, if the cooling system has enough thermal time lag, it is possible to obtain over-load skills in which the available over-load currents are managed by the proposed strategy. In case of prolonged over-load, the situation is detected and the inverter simply delivers a percentage of the nominal current to relieve the accumulated thermal stress.

Finally, a short-circuit proof strategy is introduced as a fault current limiter and voltage regulator. This short-circuit proof strategy is validated though unipolar, bipolar, tripolar and tetrapolar pure short-circuits. As the control in grid-disconnected mode is only based on voltage and current control loops, the suggested strategy only considers voltage and current, but not power. The strategy is mainly based on computing the rms value of the delivered current. Then, the current and voltage can be quickly adapted to handle the fault occurrence and clearance. This is done maintaining sinusoidal waveforms in both cases by a simple calculation of two proportional factors and the use of a filter. In this sense, the time response of protective devices will be affected minimally. The time response has been a priority in this paper, obtaining fault current limitation and voltage regulation in the order of two–three grid cycles even considering low impedance short-circuits, a superior capability compared to similar literature.

The feasibility of the proposed converter has been demonstrated not only at control level by simulation using Matlab/Simulink but also experimentally at a CENER-ATENEA four-wire microgrid.

Author Contributions: Conceptualization, D.H.-P., M.P.-G. and C.C.-A.; Software, D.H.-P., M.P.-G. and C.C.-A.; Validation, M.S., D.R. and M.A.; Formal Analysis, D.H.-P.; Investigation, D.H.-P., M.P.-G. and C.C.-A.; Methodology, D.M.-M.; Resources, M.S., D.R. and M.A.; Writing—Original Draft Preparation, D.H.-P.; Writing—Review D.H.-P., C.C.-A.

Funding: This research received no external funding.

Conflicts of Interest: The authors declare no conflict of interest.

References

1. Hatziargyriou, N. The Microgrids Concept. In *Microgrids: Architectures and Control*; IEEE: Piscataway, NJ, USA, 2014. [CrossRef]
2. Bevrani, H.; Watanabe, M.; Mitani, Y. Microgrid Control: Concepts and Classification. In *Power System Monitoring and Control*; IEEE: Piscataway, NJ, USA, 2014. [CrossRef]
3. Hartono, B.S.; Budiyanto, Y.; Setiabudy, R. Review of microgrid technology. In Proceedings of the 2013 International Conference on QiR (Quality in Research), Yogyakarta, Indonesia, 25–28 June 2013; pp. 127–132. [CrossRef]
4. Cecati, C.; Khalid, H.A.; Tinari, M.; Adinolfi, G.; Graditi, G. DC nanogrid for renewable sources with modular DC–DC LLC converter building block. *IET Power Electron.* **2017**, *10*, 536–544. [CrossRef]
5. Zubieta, L.E. Are Microgrids the Future of Energy?: DC Microgrids from Concept to Demonstration to Deployment. *IEEE Electrif. Mag.* **2016**, *4*, 37–44. [CrossRef]
6. Guerrero, J.; Vasquez, J.; Teodorescu, R. Hierarchical control of droop controlled DC and AC microgrids. In Proceedings of the 35th Annual Conference of the IEEE Industrial Electronics Society, Porto, Portugal, 3–5 November 2009.
7. Guo, F.; Wen, C.; Mao, J.; Song, Y. Distributed Secondary Voltage and Frequency Restoration Control of Droop-Controlled Inverter-Based Microgrids. *IEEE Trans. Ind. Electron.* **2015**, *62*, 4355–4364. [CrossRef]
8. Che, L.; Shahidehpour, M.; Alabdulwahab, A.; Al-Turki, Y. Hierarchical Coordination of a Community Microgrid With AC and DC Microgrids. *IEEE Trans. Smart Grid* **2015**, *6*, 3042–3051. [CrossRef]
9. Dehkordi, N.M.; Sadati, N.; Hamzeh, M. Robust tuning of transient droop gains based on Kharitonov's stability theorem in droop-controlled microgrids. *IET Gener. Transm. Distrib.* **2018**, *12*, 3495–3501. [CrossRef]
10. Han, Y.; Shen, P.; Zhao, X.; Guerrero, J.M. Control Strategies for Islanded Microgrid Using Enhanced Hierarchical Control Structure With Multiple Current-Loop Damping Schemes. *IEEE Trans. Smart Grid* **2017**, *8*, 1139–1153. [CrossRef]
11. He, J.; Li, Y.W.; Blaabjerg, F. Flexible Microgrid Power Quality Enhancement Using Adaptive Hybrid Voltage and Current Controller. *IEEE Trans. Ind. Electron.* **2014**, *61*, 2784–2794. [CrossRef]
12. Lim, K.; Choi, J. PR control based cascaded current and voltage control for seamless transfer of microgrid. In Proceedings of the 2015 IEEE 2nd International Future Energy Electronics Conference (IFEEC), Taipei, Taiwan, 1–4 November 2015; pp. 1–6. [CrossRef]
13. Rahmani-Andebili, M. Cooperative Distributed Energy Scheduling in Microgrids. In *Electric Distribution, Network Management and Control*; Springer: Singapore, 2018; pp. 235–254. [CrossRef]
14. Han, Y.; Zhang, K.; Li, H.; Coelho, E.A.A.; Guerrero, J.M. MAS-Based Distributed Coordinated Control and Optimization in Microgrid and Microgrid Clusters. A Comprehensive Overview. *IEEE Trans. Power Electron.* **2018**, *33*, 6488–6508. [CrossRef]
15. Marzband, M.; Sumper, A.; Dominguez-Garcia, J.L.; Gumara-Ferreta, R. Experimental validation of a real time energy management system for microgrids in islanded mode using a local day-ahead electricity market and MINLP. *Energy Convers. Manag.* **2013**, *76*, 314–322. [CrossRef]
16. Marzband, M.; Azarinejadian, F.; Savaghebi, M.; Pouresmaeil, E.; Guerrero, J.M.; Lightbody, G. Smart transactive energy framework in grid-connected multiple home microgrids under independent and coalition operations. *Renew. Energy* **2018**, *126*, 95–106. [CrossRef]
17. Marzband, M.; Fouladfar, M.H.; Akoredei, M.F.; Pouresmaeil, E.; Lightbody, G. Framework for Smart Transactive Energy in Home-Microgrids Considering Coalition Formation and Demand Side Management. *Sustain. Cities Soc.* **2018**, *40*, 136–154. [CrossRef]

18. Tavakoli, M.; Shokridehaki, F.; Marzband, M.; Godinac, R.; Pouresmaeil, E. A two stage hierarchical control approach for the optimal energy management in commercial building microgrids based on local wind power and PEVs. *Sustain. Cities Soc.* **2018**, *41*, 332–340. [CrossRef]

19. Hatziargyriou, N. Microgrid Protection. In *Microgrids: Architectures and Control*; IEEE: Piscataway, NJ, USA, 2014. [CrossRef]

20. Sharkh, S.M.; Abu-Sara, M.A.; Orfanoudakis, G.I.; Hussain, B. Microgrid Protection. In *Power Electronic Converters for Microgrids*; IEEE: Piscataway, NJ, USA, 2014. [CrossRef]

21. Carpio-Huayllas, T.E.D.; Ramos, D.S.; Vasquez-Arnez, R.L. Microgrid transition to islanded modes: Conceptual background and simulation procedures aimed at assessing its dynamic performance. In Proceedings of the Transmission and Distribution Conference and Exposition (T D), Orlando, FL, USA, 7–10 May 2012; pp. 1–6. [CrossRef]

22. Heredero-Peris, D.; Chillón-Antón, C.; Pagès-Giménez, M.; Gross, G.; Montesinos-Miracle, D. Implementation of grid-connected to/from off-grid transference for micro-grid inverters. In Proceedings of the IECON 2013—39th Annual Conference of the IEEE Industrial Electronics Society, Vienna, Austria, 10–13 November 2013; pp. 840–845. [CrossRef]

23. Liu, Z.; Liu, J. Unified control based seamless transfer of microgrids. In Proceedings of the 2015 9th International Conference on Power Electronics and ECCE Asia (ICPE-ECCE Asia), Seoul, Korea, 1–5 June 2015; pp. 1477–1482. [CrossRef]

24. Shen, Z.J. Ultrafast Solid-State Circuit Breakers: Protecting Converter-Based ac and dc Microgrids Against Short Circuit Faults [Technology Leaders]. *IEEE Electrif. Mag.* **2016**, *4*, 72–70. [CrossRef]

25. Laaksonen, H.J. Protection Principles for Future Microgrids. *IEEE Trans. Power Electron.* **2010**, *25*, 2910–2918. [CrossRef]

26. Zhang, J.; Wu, P.; Hong, J. Control strategy of microgrid inverter operation in Grid-connected and Grid-disconnected modes. In Proceedings of the International Conference on Electric Information and Control Engineering (ICEICE), Wuhan, China, 25–27 March 2011; pp. 1257–1260. [CrossRef]

27. Shan, W.C.; Lin, L.X.; Li, G.; Wei, L.Y. A seamless operation mode transition control strategy for a microgrid based on master–slave control. In Proceedings of the 2012 31st Chinese Control Conference, Hefei, China, 25–27 July 2012; pp. 6768–6775.

28. Brabandere, K.D. Voltage and Frequency Droop Control in Low Voltage Grids by Distributed Generators with Inverter Front-End. Ph.D. Thesis, Katholieke Universiteit Leuven, Leuven, Belgium, 2006.

29. Wessels, C.; Dannehl, J.; Fuchs, F. Active damping of LCL-filter resonance based on virtual resistor for PWM rectifiers; stability analysis with different filter parameters. In Proceedings of the 2008 IEEE Power Electronics Specialists Conference, Rhodes, Greece, 15–19 June 2008; pp. 3532–3538. [CrossRef]

30. Kim, J.; Guerrero, J.; Rodríguez, P.; Teodorescu, R.; Nam, K. Mode Adaptive Droop Control with Virtual Output Impedances for an Inverter-Based Flexible AC Microgrid. *IEEE Trans. Power Electron.* **2011**, *26*, 689–701. [CrossRef]

31. Guo, X.; Lu, Z.; Wang, B.; Sun, X.; Wang, L.; Guerrero, J. Dynamic Phasors-Based Modeling and Stability Analysis of Droop-Controlled Inverters for Microgrid Applications. *IEEE Trans. Smart Grid* **2014**, *5*, 2980–2987. [CrossRef]

32. Bayindir, R.; Irmak, E.; Issi, F.; Guler, N. Short-circuit fault analysis on microgrid. In Proceedings of the 2015 International Conference on Renewable Energy Research and Applications (ICRERA), Palermo, Italy, 22–25 November 2015; pp. 1248–1252. [CrossRef]

33. Almutairy, I. A review of coordination strategies and techniques for overcoming challenges to microgrid protection. In Proceedings of the 2016 Saudi Arabia Smart Grid (SASG), Jeddah, Saudi Arabia, 6–8 December 2016; pp. 1–4. [CrossRef]

34. Lai, X.; Liu, F.; Deng, K.; Gao, Q.; Zha, X. A short-circuit current calculation method for low-voltage DC microgrid. In Proceedings of the 2014 International Power Electronics and Application Conference and Exposition, Shanghai, China, 5–8 November 2014; pp. 365–371. [CrossRef]

35. Ouaida, R.; Berthou, M.; Tournier, D.; Depalma, J. State of art of current and future technologies in current limiting devices. In Proceedings of the 2015 IEEE First International Conference on DC Microgrids (ICDCM), Atlanta, GA, USA, 7–10 June 2015; pp. 175–180. [CrossRef]

36. Balasreedharan, S.S.; Thangavel, S. An adaptive fault identification scheme for DC microgrid using event based classification. In Proceedings of the 2016 3rd International Conference on Advanced Computing and Communication Systems (ICACCS), Coimbatore, India, 22–23 January 2016; Volume 1, pp. 1–7. [CrossRef]

37. Cairoli, P.; Rodrigues, R.; Zheng, H. Fault current limiting power converters for protection of DC microgrids. In Proceedings of the SoutheastCon 2017, Charlotte, NC, USA, 30 March–2 April 2017; pp. 1–7. [CrossRef]

38. Pei, X.; Chen, Z.; Wang, S.; Kang, Y. Overcurrent protection for inverter-based distributed generation system. In Proceedings of the 2015 IEEE Energy Conversion Congress and Exposition (ECCE), Montreal, QC, Canada, 20–24 September 2015; pp. 2328–2332. [CrossRef]

39. Babqi, A.J.; Etemadi, A.H. MPC-based microgrid control with supplementary fault current limitation and smooth transition mechanisms. *IET Gener. Transm. Distrib.* **2017**, *11*, 2164–2172. [CrossRef]

40. Beheshtaein, S.; Savaghebi, M.; Guerrero, J.M.; Cuzner, R.; Vasquez, J.C. A secondary-control based fault current limiter for four-wire three phase inverter-interfaced DGs. In Proceedings of the IECON 2017—43rd Annual Conference of the IEEE Industrial Electronics Society, Beijing, China, 29 October–1 November 2017; pp. 2363–2368. [CrossRef]

41. Rahmatian, M.; Sanjari, M.J.; Gholami, M.; Gharehpetian, G.B. Optimal control of distribution line series compensator in microgrid considering fault current limitation function. In Proceedings of the 2012 17th Conference on Electrical Power Distribution, Tehran, Iran, 2–3 May 2012; pp. 1–5.

42. Sadeghkhani, I.; Golshan, M.E.H.; Mehrizi-Sani, A.; Guerrero, J.M. Low-voltage ride-through of a droop-based three-phase four-wire grid-connected microgrid. *IET Gener. Transm. Distrib.* **2018**, *12*, 1906–1914. [CrossRef]

43. Li, Y.; Vilathgamuwa, D.M.; Loh, P.C. Microgrid power quality enhancement using a three-phase four-wire grid-interfacing compensator. *IEEE Trans. Ind. Appl.* **2005**, *41*, 1707–1719. [CrossRef]

44. Kim, G.H.; Hwang, C.; Jeon, J.H.; Byeon, G.S.; Ahn, J.B.; Jo, C.H. Characteristic analysis of three-phase four-leg inverter based load unbalance compensator for stand-alone microgrid. In Proceedings of the 2015 9th International Conference on Power Electronics and ECCE Asia (ICPE-ECCE Asia), Seoul, Korea, 1–5 June 2015; pp. 1491–1496. [CrossRef]

45. Liu, Y.-H.; Yang, J.; Li, H.; Wang, C. The research of three phase four wire active power filter on small independent micro-grid. In Proceedings of the 2016 China International Conference on Electricity Distribution (CICED), Xi'an, China, 10–13 August 2016; pp. 1–4. [CrossRef]

46. Lee, W.; Han, B.M.; Cha, H. Battery ripple current reduction in a three-phase interleaved dc-dc converter for 5 kW battery charger. In Proceedings of the 2011 IEEE Energy Conversion Congress and Exposition, Phoenix, AZ, USA, 17–22 September 2011; pp. 3535–3540. [CrossRef]

47. Khosroshahi, A.; Abapour, M.; Sabahi, M. Reliability Evaluation of Conventional and Interleaved DC–DC Boost Converters. *IEEE Trans. Power Electron.* **2015**, *30*, 5821–5828. [CrossRef]

48. Llonch-Masachs, M.; Heredero-Peris, D.; Montesinos-Miracle, D.; Rull-Duran, J. Understanding the three and four-leg inverter Space Vector. In Proceedings of the 2016 18th European Conference on Power Electronics and Applications (EPE'16 ECCE Europe), Karlsruhe, Germany, 5–9 September 2016; pp. 1–10. [CrossRef]

49. Heredero-Peris, D.; Pagès-Giménez, M.; Montesinos-Miracle, D. Inverter design for four-wire microgrids. In Proceedings of the 2015 17th European Conference on Power Electronics and Applications (EPE'15 ECCE-Europe), Geneva, Switzerland, 8–10 September 2015; pp. 1–10. [CrossRef]

50. Rivas, H.A.; Bergas, J. Frequency Determination in a Single-Phase Voltage Signal using Adaptive Notch Filters. In Proceedings of the 2007 9th International Conference on Electrical Power Quality and Utilisation, Barcelona, Spain, 9–11 October 2007; pp. 1–7. [CrossRef]

51. Yepes, A.G. Digital Resonant Current Controllers for Voltage Source Converters. Ph.D. Thesis, Universidad de Vigo, Pontevedra, Spain, 2011.

52. Rodríguez, F.; Bueno, E.; Aredes, M.; Rolim, L.; Neves, F.; Cavalcanti, M. Discrete-time implementation of second order generalized integrators for grid converters. In Proceedings of the 2008 34th Annual Conference of IEEE Industrial Electronics, Orlando, FL, USA, 10–13 November 2008; pp. 176–181.

53. Balaras, C.A. The role of thermal mass on the cooling load of buildings. An overview of computational methods. *Energy Build.* **1996**, *24*, 1–10. [CrossRef]

54. Mjallal, I.; Farhat, H.; Hammoud, M.; Ali, S.; Assi, I. Improving the Cooling Efficiency of Heat Sinks through the Use of Different Types of Phase Change Materials. *Technologies* **2018**, *6*, 5. [CrossRef]

55. *ITI (CBEMA) Curve Application Note*; Techreport; Information Technology Industry Council: Washington, DC, USA, 2000.

applied
sciences

MDPI

Article

Research on Operation–Planning Double-Layer Optimization Design Method for Multi-Energy Microgrid Considering Reliability

Shaoyun Ge, Jifeng Li, Hong Liu *, Hao Sun and Yiran Wang

Key Laboratory of Smart Grid, Tianjin University, Tianjin 300000, China; syge@tju.edu.cn (S.G.); lijifeng2014@163.com (J.L.); sunhao1993@tju.edu.cn (H.S.); Wang_Yiran94@163.com (Y.W.)
* Correspondence: liuhong@tju.edu.cn

Received: 13 September 2018; Accepted: 22 October 2018; Published: 25 October 2018

Featured Application: The contents of this article can provide guidance for the construction of integrated energy system.

Abstract: A multi-energy microgrid has multiple terminal resources and multiple distributed components for energy production, conversion, and storage. By using this grid, an interconnected network with optimized multiple energy sources can be formed. This type of grid can minimize energy waste while laying the critical foundation for an energy Internet. The multi-energy microgrid must be formed properly to ensure multi-energy coupling and complement. However, critical technologies (e.g., reliability assessment) and configuration planning methods now need further research. In this study, a novel method for the reliability evaluation of a multi-energy supply is proposed, and an operation–planning double-layer optimization design method is investigated that considers reliability. On that basis, the effects of different configuration schemes on economy and reliability are quantitatively analyzed. First, the coupling relationship between multi-energy carriers in a typical multi-energy microgrid is analyzed; subsequently, the energy efficiency and economical models of the key equipment in the grid system are determined. Monte Carlo simulation and the Failure Mode and Effect Analysis (FMEA) method are applied to evaluate the reliability with sorted indicators. A double-layer optimization model is built for a multi-energy microgrid with the optimal configuration. The impact of configuration on the reliability and economical performance of the microgrid system is quantitatively analyzed based on actual calculations. The results obtained here are relative to the capacity, configuration, operation, and energy supply reliability of the multi-energy microgrid, and may serve as the feasible guidelines for future integrated energy systems.

Keywords: multi-energy microgrid; energy hub; optimal planning; reliability

1. Introduction

Energy lays the foundation for human existence. It is also the source for social and economic development. In recent years, there has been increasing conflict between energy supply and energy demand, as well as rising environmental concerns. New technologies are urgently needed to improve energy efficiency while ensuring an effective clean energy supply [1]. With the construction of today's energy Internet, there is also a demand of synergistic multi-energy (power, gas, and heat) supply systems to replace the independent operation modes of the traditional energy supply system. The multi-energy microgrid, which is a key node in the energy Internet, has become a popular research subject due to its flexible operation modes and effective energy use forms. New planning and reliability evaluation methods for integrated energy systems are important in this regard.

Reference [2] includes the first energy hub model, conceptually illustrating the multi-energy microgrid. This model reflects the complicated relationship of the energy hub from the perspective of energy flow via an energy conversion matrix, and delineates the relations between each energy subsystem in the hub. Reference [3] further expanded the integrated energy system concept and analyzed the characteristics of such systems per the coupling components and synergy of different energy vectors. The emergence of the electric microgrid concept [4–6] has brought about new multi-energy microgrids with flexible operation modes, having evolved from the traditional distributed energy supply and storage systems. These new systems can ensure the effective accommodation of renewable energy with the mutual support of utility energy grids. The continuous development of microgrid technology will make multi-energy microgrid become an important node of the energy Internet and an effective comprehensive energy utilization model.

An accurate reliability evaluation is the basis of any effective planning scheme. The strengthening of coupling characteristics between multi-energy systems affects the reliability of the energy supply. Due to the coupling between energy carriers, there are multiple energy supply points on the demand side, which enhances the reliability of the energy supply. However, as energy supply systems couple with each other, a fault in any one system affects the entire system. Previous researchers have explored various reliability evaluation methods. Analytical and simulation methods [7,8] are commonly employed to power systems, and may also apply to the assessment of microgrid reliability [9,10]. However, there has been relatively little research on integrated energy supply reliability evaluation. Some references have extended the range of reliability assessment by adding distributed generation to traditional distribution networks. Here are the examples. Reference [11] employed scenario reduction techniques to assess uncertainties in distributed generation and load in distribution networks; it also explored various strategies to protect against failure in the distributed network. References [12,13] evaluated the reliability of active distribution networks through analytic and simulation methods, respectively. Reference [12] analyzed the energy hub operation mode and built mathematical models for reliability evaluation that apply to different respective operation strategies. Reference [13] developed a two-hierarchy smart agent model to evaluate active distribution network reliability with Monte Carlo simulations. On that basis, some references further integrated various energy carriers and proposed some static reliability evaluation methods. References [14,15] built a state space for energy transmission by analyzing the connection between different energies in multi-energy network hubs. They also performed static reliability evaluation of the systems mathematically. Reference [16] built a state model of power grid and gas network components and a reliability evaluation model of an integrated system via the Monte Carlo simulation method. It also performed reliability evaluations at different time periods. The above references have laid a solid basis for the reliability evaluation method of multi-energy supply in this study. However, the coupling characteristics of multi-energy and the simulation of time scale remain insufficient.

One of the goals of reliability evaluation is to provide the corresponding guidance for the reasonable planning of the system, and an effective and reasonable planning method lays a solid foundation for the construction of the energy Internet in the future. At present, there is a certain research foundation for the collaborative planning method of a multi-energy network. In terms of planning methods, Reference [17], e.g., assessed the impact of different equipment configuration schemes on the reliability and economical performances of integrated microgrid systems. However, this study has only conducted concrete research on the electric power microgrid. Reference [18] proposed a multi-energy system planning method and analyzed the impact of a reliability constraint on the selection of system planning schemes. However, the paper took the simple index as the reliability constraint without systematically introducing the multi-energy comprehensive reliability evaluation method. Reference [19] took Masirah Island in Oman as an example to analyze the technical and economic viability of a hybrid energy system; different scenarios were established based on different configurations of equipment. On this basis, the impacts of different scenarios were considered in the hybrid system optimization. This study laid a solid empirical basis for subsequent planning implementation. In accordance with the principle of the Bayesian filter structure, Reference [20] proposed a state estimation and stabilization algorithm for smart grid and further designed a semidefinite programming-based optimal feedback controller. This study has built a good foundation for further analysis, evaluation, and adjustment of the state of the system. Besides, some references have also studied the expansion planning methods of energy hubs. Reference [21] proposed a planning and configuration method for an energy hub with multiple energy systems, and further assessed the performance of the planning program in terms of energy efficiency and emissions.

Despite the wealth of valuable information on this subject, there is much work to be done to ensure the safe and practical application of multi-energy supply systems. The above researchers primarily emphasized the single energy network level represented by electric energy, and neglected the impacts of peak and valley differences in various energy load demands; thus, their results do not reflect the specific requirements for integrated energy synergistic planning. Mathematical analysis methods are usually employed to analyze reliability, whereas these methods do not reflect the characteristics of the unit and user demand in the timing series. Researchers also tend to ignore the respective energy grades of different energy sources, thereby not securing feasible energy supply/storage priority and load reduction strategies. Traditional planning methods are tailored to operation, in which the optimization goal is the minimal comprehensive cost formed by the operation cost, startup cost, and fuel cost of the system; the randomness of unit failure and the reliability of the power system are typically not considered. The economical factors under normal operation as well as any possible risks of the system must be properly accounted in planning the configuration of a multi-energy microgrid.

Given the above problems, this study selected a multi-energy microgrid as its research target and employed the energy hub model to analyze the coupling relationship between multi-energy networks. Subsequently, the energy efficiency and economical models of the key equipment in the multi-energy microgrid system were proposed in this study. Besides, through Monte Carlo simulation and the Failure Mode and Effect Analysis (FMEA) method, the paper raised a novel reliability assessment method for the energy supply of multiple energy carriers and sorted reliability assessment indicators. On this basis, a double-layer planning model with optimal configuration, operation, and reliability consideration for the multi-energy microgrid was built. Finally, based on the actual calculations, this study conducted a quantitative analysis on the impact of different configuration schemes on the reliability and economical performances of the microgrid system, and proposed the application of reliability in the maintenance of key equipment.

The major contributions of this study are summarized as follows: (1) a novel multi-energy integrated reliability evaluation method was proposed; (2) the reliability factor was integrated into the double-layer planning model of a multi-energy microgrid; (3) the impacts of different equipment configuration schemes on the planning economy and reliability were analyzed quantitatively through practical cases, and the application of reliability in practical maintenance was further analyzed by cases. This study considered the reliability of microgrid system energy supply during the optimization configuration process, and proposed the reliability assessment indicators of the microgrid system energy supply. It combined capacity optimization configuration with operation strategies, and designed the assembly style and capacity in the microgrid system to make the energy supply reliability meet the expectations and minimize the cost in the planning period.

2. Multi-Energy Microgrid Structure

The multi-energy microgrid is an energy system that runs autonomously. The system consists of energy management devices, distributed renewable energy sources, energy storage devices, energy conversion devices, and energy loads; structurally, the microgrid is composed of energy input, conversion, and storage, as well as output components. In this study, a multi-energy microgrid was built based on the energy hub model, which consisting of a combined cold heat and power (CCHP) system, a gas heat pump (GHP), a distributed photovoltaic (PV), central air conditioning (CAC), electric storage (ES), and heat storage (HS) components with power, gas, cold, and heat energies. The structure of the multi-energy microgrid is shown in Figure 1.

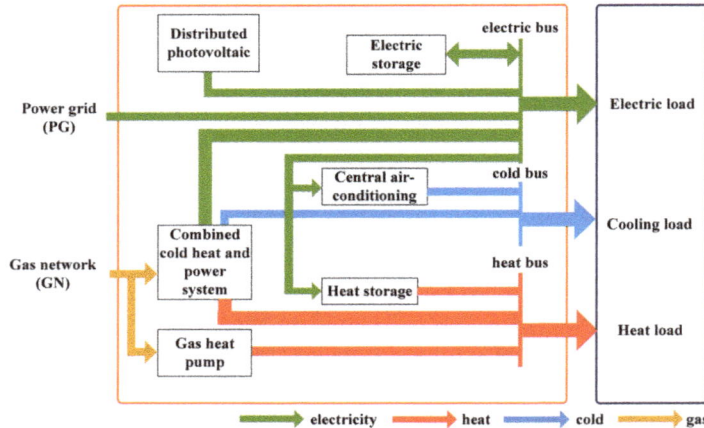

Figure 1. Structure of multi-energy microgrid.

The multi-energy microgrid contains various equipment units that couple with each other to meet the needs of different energy loads. Power loads are supplied by the CCHP system and PV. The ES or power grid (PG) make up for insufficient outputs; heat loads are supplied by the CCHP and GHP. HS devices make up for insufficient outputs; cold loads are supplied by the CCHP and CAC. In the case of failure, different forms of energy give priority to the supply of energy loads and supplement of energy storage devices of the same kind. Extra energy can be employed for conversion or backup to reflect the differences among energy grades.

The multi-energy microgrid is an independent and controllable basic unit of the energy Internet at the user terminal. Its flexible operation mode can reduce the energy consumption during transmission and realize the effective consumption of renewable energy sources. The reliability of two typical microgrid system operation modes—the grid-connected mode and the electric-isolated mode—was analyzed to fully consider the reliability factors. Subsequently, a novel planning method for the electric-isolated mode system was proposed. The main assumptions regarding the multi-energy microgrid model are as follows.

(1) Energy distribution networks are placed into a single bus radiation network structure with a certain level of isolation between the devices in the multi-energy microgrid.
(2) Different types of energy loads are intensively provided by relevant devices.
(3) Failure in any one-unit device occurs independently; only single failure is considered [22].

3. Equipment Efficiency and Economical Models

3.1. Gas-Fired CCHP System

The CCHP system, the key unit of the microgrid system, provides energy for internal equipment (e.g., absorption chiller and heat exchanger) and the cooling, heating, and power loads of the microgrid system. Configuration planning of the multi-energy microgrid centers around the CCHP output and fuel consumption. The efficiency model of the gas-fired CCHP system is [23]:

$$
\begin{cases}
Q_{MT} = P_{output}^{CCHP}(1 - \eta_{MT}^{CCHP} - \eta_1^{CCHP})/\eta_{MT}^{CCHP} \\
Q_{h0} = Q_{MT}\eta_{rec}^{CCHP}K_{h0} \\
Q_{c0} = Q_{MT}\eta_{rec}^{CCHP}K_{c0} \\
\eta_{rec}^{CCHP} = \frac{T_1 - T_2}{T_1 - T_0} \\
V_{MT}^{CCHP} = (\sum P_{output}^{CCHP}\Delta t)/(\eta_{MT}^{CCHP}L)
\end{cases}
\tag{1}
$$

where Q_{MT} is the exhaust residual heat of the CCHP system; P_{output}^{CCHP} is the output power of the CCHP system; η_{MT}^{CCHP} is the efficiency of the gas turbine, which is set as 75%; η_1^{CCHP} is the heat dissipation coefficient of the CCHP system, which is set as 25%; Q_{h0} and Q_{c0} are the heating or refrigerating capacity provided by the residual heat from the gas turbine, respectively; η_{rec}^{CCHP} is the recovery efficiency of waste heat from flue gas; K_{h0} and K_{c0} are the heating or refrigerating coefficient of bromine cooler, which are set as 1.2 and 0.95, respectively; T_1 and T_2 are the environmental coefficients, which are set as 573.15 K and 423.25 K, respectively; V_{MT}^{CCHP} is the amount of natural gas consumed by the CCHP system; L is the low calorific value of natural gas, which is set as 9.7 kW·h/m³; and Δt is the unit run time.

Under Equation (1), the CCHP uses waste heat for heating or cooling. If there are remaining heating and cooling loads to be satisfied, extra gas consumption is necessary during the warming up period [23]; that is:

$$
V_{h1}^{CCHP} = \frac{\sum (Q_h - Q_{h0})\Delta t_h}{K_{h0}\eta_{in}^{CCHP}L}
\tag{2}
$$

$$
V_{c1}^{CCHP} = \frac{\sum (Q_c - Q_{c0})\Delta t_c}{K_{c0}\eta_{in}^{CCHP}L}
\tag{3}
$$

where V_{h1}^{CCHP} and V_{c1}^{CCHP} are the amount of natural gas that is needed for heating or refrigeration in the CCHP system; Q_h and Q_c are the heat or cold load to be satisfied by the CCHP system; η_{in}^{CCHP} is the operation efficiency of the CCHP for supplementary combustion, which is set as 75%; and Δt_h and Δt_c are operation time of gas turbine in heating or cooling, respectively.

Apart from the inherent installation cost, the fuel cost of the CCHP system can be calculated as follows:

$$
C_{fuel}^{CCHP} = (w^{gas}/L)\sum_t (P_{output}^{CCHP}/\eta_{MT}^{CCHP})
\tag{4}
$$

where C_{fuel}^{CCHP} is the fuel cost of CCHP system; and w^{gas} is natural gas price.

The operation and maintenance costs of the CCHP can be calculated as follows:

$$
C_{opma}^{CCHP} = K_{MT}^{CCHP}\sum_t P_{output}^{CCHP}
\tag{5}
$$

where C_{opma}^{CCHP} is the operation and maintenance costs of the CCHP system; and K_{MT}^{CCHP} is the proportional coefficient of the CCHP's operation and maintenance costs.

3.2. Energy Conversion Equipment

The multi-energy microgrid contains several pieces of energy conversion equipment, including the GHP and CAC. Combined with the energy hub model, the efficiency output model of the energy conversion equipment is [2]:

$$
P_b(t) = C_{ab}P_a(t)
\tag{6}
$$

where C_{ab} is the energy conversion coefficient.

Besides the equipment installation cost, operation and maintenance costs also must be considered during the planning phase. The operation and maintenance costs of the energy conversion equipment can be calculated as follows [24]:

$$C_{opma}^{Trans}(t) = \Delta t \sum_{m=1}^{N_{Trans}} K_{MT}^m \left| P_{output}^m(t) \right| \tag{7}$$

where N_{Trans} is the total number of energy conversion equipment units.

3.3. PV Power Generation System

The output model of the PV system is [25]:

$$P_{PV}(t) = f_{PV} P_{PV,rated} \frac{A(t)}{A_S} \left[1 + \alpha_p (TEM(t) - TEM_{STC}) \right] \tag{8}$$

where f_{PV} is the power reduction factor of PV, which is set as 0.9; $P_{PV,rated}$ is the rated power of PV; $A(t)$ is the actual irradiance on the PV current tilt plane; A_S is the irradiance under standard test conditions, which is set as 1 kW/m^2; α_p is the power temperature coefficient, which is set as $-0.47\%/^\circ C$; $TEM(t)$ is the surface temperature of PV; and TEM_{STC} is the temperature of PV under standard test conditions, which is set as 25 $^\circ$C in this study.

Unlike other components in the multi-energy microgrid, the PV system does not need primary energy. Thus, no fuel cost is needed. It is much less expensive to operate PV than it is to run other equipment. By fully considering the PV construction cost, maintenance cost, labor cost, price increase indices, government subsidies, the actual service life of the PV system, the net PV residual value, and the unit conversion cost of PV, power generation is calculated to simplify the following economical calculations. The conversion cost is as follows:

$$C_{output}^{PV} = \frac{N_{PV} P_{PV,rated} c_{PV0}(1-\rho) + \sum_{Y_{PV}=1}^{N_{Y,PV}} f(Y)}{N_{Y,PV} N_{PV} T_{PV,max} M_{PV}} - B_{subsidy} \tag{9}$$

where N_{PV} is the number of PVs; $c_{PV,0}$ is the unit capacity investment in PV construction; ρ is the net residual rate, which is set as 3%; $f(Y)$ is the net residual value of PV; M_{PV} is the installation capacity of PV; and $B_{subsidy}$ is the national subsidy for energy-saving and emission reduction, which is set as 0.1 ¥/kW·h.

The net residual value of PV is:

$$f(Y) = (1 + \beta_{PV})^Y C_{oper}^{PV} \tag{10}$$

where β_{PV} is the price rise coefficient, which is set as 5%; and Y is the planning life cycle, which is set as 12a.

3.4. ES Devices

The lead–acid batteries (LB) are employed as electric storage devices for the microgrid system that is modeled in this study. Compared with other power storage technologies, the LB features higher charging efficiency and energy density, and it is not constrained by space, so it is a better fit for the microgrid system. The dynamic model is [26]:

$$SOC(t) = (1 - \delta)SOC(t-1) + P_C^{ES}(t)\Delta t \eta_C^{ES}/E_{SOC} - \frac{P_D^{ES}(t)\Delta t}{E_{SOC}\eta_D^{ES}} \tag{11}$$

where $SOC(t)$ is the electric quantity of ES at time interval t; δ is the ES self-discharge rate, which is set as 0.5%/h; $P_C^{ES}(t)$ and $P_D^{ES}(t)$ are the charging and discharging power of ES at time interval t, respectively; η_C^{ES} and η_D^{ES} are the charging or discharging efficiency of ES, which are set as 90% and 90%, respectively; and E_{SOC} is the rated capacity of ES.

LB has a low self-discharge rate, so self-discharge loss is ignored. The cost lost in ES during operation mainly includes the costs of life loss C_{life}^{ES} and power transmission loss C_{trans}^{ES}. The cost of life loss in the power storage device can be calculated as follows:

$$C_{life}^{ES} = \frac{\mu^{ES} C_{invest}^{ES}}{N(|ES(t-1) - ES(t)|)} \tag{12}$$

where μ^{ES} is the regulating coefficient of ES, which is set as 1; C_{invest}^{ES} is the one-time ES purchase cost; $N(\cdot)$ is the discharge depth function, which is typically a four-order function that denotes the relationship between cycle life and discharge depth [23]; $ES(t)$ is the power energy proportion of the current LB, which can be calculated as follows:

$$ES(t) = \frac{SOC(t)}{E_{SOC}} \tag{13}$$

The power transmission loss C_{trans}^{ES} of ES devices results from incomplete transmission. It can be expressed as follows:

$$C_{trans}^{ES}(t) = c_{trans}^{ES}[(1 - \eta_C^{ES})P_C^{ES}(t) + (1/\eta_D^{ES} - 1)P_D^{ES}(t)] \tag{14}$$

When the multi-energy microgrid system runs in grid-connected mode, ES devices serve to stabilize the fluctuation of renewable energy and improve the economical performance of the system. The charging and discharging states of ES devices are associated with the output of the relevant unit and the electricity pricing mechanism. When the microgrid system runs in electric-isolated mode, the ES devices provide a significant backup energy supply. When something goes awry in the microgrid system, extra power energy will be supplied first to the ES devices and similar types of loads. Energy conversion will be possible if extra energy sources remain.

3.5. HS Devices

The regenerative electric boiler is selected as the HS device in the system model. It not only stores heat; it also helps absorb renewable energy sources. The dynamic model of HS devices is [23]:

$$H_{HS}(t) = (1 - k_{LOSS})H_{HS}(t-1) + Q_C^{HS}(t)\Delta t \eta_C^{HS} - \frac{Q_D^{HS}(t)\Delta t}{\eta_D^{HS}} \tag{15}$$

where $H_{HS}(t)$ is the heat of HS at time interval t; k_{LOSS} is the heat dissipation rate of HS, which is set as 1%/h; η_C^{HS} and η_D^{HS} are the charging and discharging efficiency of HS, which are set as 90% and 90%, respectively.

The circular life loss of the HS devices is smaller than that of the ES devices, and is not considered in this study. The operating cost of the HS device mainly includes the idle cooling cost and heat transmission loss cost. The cost of HS device loss can be calculated as follows:

$$C_{oper}^{HS}(t) = c_{h0}^{HS}[k_{LOSS}H_{HS}(t-1) + (1 - \eta_C^{HS})Q_C^{HS}(t)\Delta t + (1/\eta_D^{HS} - 1)Q_D^{HS}(t)\Delta t] \tag{16}$$

4. Multi-Energy Microgrid Reliability Assessment

Reliability lays the foundation for any ideal planning scheme. As mentioned above, FMEA and Monte Carlo simulation were combined to remedy the problems inherent to extant reliability assessment methods. FMEA was employed to analyze the logical relationship of energy supply in the microgrid system, and quantify the potential effects of unit fault on different types of energy supply. Thus, their various coupling features can be reflected. Monte Carlo simulation is a reliability assessment method that can be applied to the multi-energy supply of the microgrid to simulate the timing sequence features of the equipment units and load demands in the system.

Taking grid-connected operation mode as the example, the reliability status of different energy types (power, heat, and cold) can be calculated as follows:

$$\begin{aligned}
R_{\text{elec}} &= f(S_{\text{CCHP}}, S_{\text{PV}}, S_{\text{ES}}, S_{\text{PG}}, S_{\text{GN}}) \\
R_{\text{heat}} &= f(S_{\text{CCHP}}, S_{\text{GHP}}, S_{\text{HS}}, S_{\text{GN}}) \\
R_{\text{cold}} &= f(S_{\text{CCHP}}, S_{\text{CAC}}, S_{\text{GN}})
\end{aligned} \tag{17}$$

where R is the reliability of an energy supply; $f(\cdot)$ is the calculation function of energy supply reliability; and $S(\cdot)$ is the state of equipment.

Based on the multi-energy microgrid shown in Figure 1, the failures in the CCHP, GHP, CAC, PV, ES, PG, and gas network (GN) of the microgrid system were analyzed. Also, this study divided one year into 8760 one-hour time periods. In each period, the equipment status parameters were kept unchanged, and the instantaneous quantity was recorded at specific time points.

4.1. State Model of Equipment Units

In the microgrid system, the output of units (e.g., PV or CCHP) and different energy load demands have robust time-sequential characteristics. It is challenging to model these characteristics mathematically, which affects the feasibility of any reliability evaluation. A state model of equipment units must be built to simulate the equipment unit states via the Monte Carlo method. Subsequently, the time-sequential characteristics in the microgrid system can be modeled appropriately.

The Markov two-state models [8] were employed to describe the equipment units of the microgrid system. The normal state of a unit is presented in an exponential distribution. The duration from the unit's normal working state to failure is:

$$T_{f,k} = -(\frac{1}{\lambda_k}) \ln u_k \, k \in [1, 2, \ldots \ldots m] \tag{18}$$

where λ_k is the failure rate of kth equipment unit; and u_k represents the random numbers of uniform distribution in [0,1] interval.

All of the equipment units are repairable. The failure duration is also an exponential distribution:

$$T_{r,k} = -(\frac{1}{\mu_k}) \ln u_k \, k \in [1, 2, \ldots \ldots m] \tag{19}$$

where μ_k is the repair rate of kth equipment unit.

4.2. Reliability Evaluation Indexes

This study selected the expectation of energy supplied (EES), loss of energy expectation (LOEE), and system average interruption duration index (SAIDI) as the indices to evaluate the reliability of different energy types in the multi-energy microgrid. The multi-energy reliability economic indicator was built by combining energy prices.

(1) EES

This index refers to the probability that a certain energy type reaches the user demand side in a certain period. A higher EES suggests that the energy can meet the demand in the sampling period with high reliability. The specific solution is as follows:

The energy conversion matrix [2] in the energy hub model is extended to the energy transmission power matrix C. Matrix C includes the unit supply matrix C_M and adjustable resource matrix C_S. The matrix expression forms are shown in Equations (20) and (21). The matrix rows represent different run-time scenarios (including normal operation and failure scenarios); the matrix columns represent the supplies of power, heat, and cold energy types. The unit supply matrix C_M refers to the energy output of the unit following the different energy demands in a certain period; the adjustable resource matrix C_S refers to the energy provided by backup resources or a utility grid to the microgrid after a certain period. Specific values are associated with specific operation strategies. The energy transmission matrix of the ith scenario in the sampling period is: $C_i = C_{M,i} + C_{S,i}$.

$$
C_M = \begin{bmatrix}
P_{output}^{CCHP} + P_{output}^{PV} & P_{output}^{CCHP} + P_{output}^{GHP} & P_{output}^{CCHP} + P_{output}^{CAC} \\
P_{output}^{PV} & P_{output}^{GHP} & P_{output}^{CAC} \\
P_{output}^{CCHP} + P_{output}^{PV} & P_{output}^{CCHP} + P_{output}^{GHP} & P_{output}^{CCHP} \\
P_{output}^{CCHP} + P_{output}^{PV} & P_{output}^{CCHP} & P_{output}^{CCHP} + P_{output}^{CAC} \\
P_{output}^{CCHP} + P_{output}^{PV} & P_{output}^{CCHP} + P_{output}^{GHP} & P_{output}^{CCHP} + P_{output}^{CAC} \\
P_{output}^{PV} & 0 & P_{output}^{CAC} \\
P_{output}^{CCHP} & P_{output}^{CCHP} + P_{output}^{GHP} & P_{output}^{CCHP} + P_{output}^{CAC} \\
P_{output}^{CCHP} + P_{output}^{PV} & P_{output}^{CCHP} + P_{output}^{GHP} & P_{output}^{CCHP} + P_{output}^{CAC}
\end{bmatrix} \tag{20}
$$

$$
C_S = \begin{bmatrix}
P_{output}^{PG} + P_{output}^{ES} & P_{output}^{HS} & 0 \\
P_{output}^{PG} + P_{output}^{ES} & P_{output}^{HS} & 0 \\
P_{output}^{PG} + P_{output}^{ES} & P_{output}^{HS} & 0 \\
P_{output}^{PG} + P_{output}^{ES} & P_{output}^{HS} & 0 \\
P_{output}^{ES} & P_{output}^{HS} & 0 \\
P_{output}^{PG} + P_{output}^{ES} & P_{output}^{HS} & 0 \\
P_{output}^{PG} + P_{output}^{ES} & P_{output}^{HS} & 0 \\
P_{output}^{PG} & P_{output}^{HS} & 0
\end{bmatrix} \tag{21}
$$

The load energy demand matrix $L = [P_{Load}^{elec}, P_{Load}^{heat}, P_{Load}^{cold}]$ reflects the demand for power, heat, and cold energies in the sampling period. The supply and demand comparison matrix Q can be built accordingly. The elements in matrix Q are Boolean variables. The elements $Q_{j,i,k}$ ($j = 1, 2, \ldots N; i = 1, \ldots \ldots, 8; k = e, h, c$) in the matrix represent the j-th sampling, i-th scene obtained, and discrimination of the k-th energy source. The distinguishing method of power energy, e.g., in the matrix is as follows:

$$
Q_{j,i,e} = \begin{cases} 0 & P_{Load}^{elec} > C_{M,i,e} + C_{S,i,e} \\ 1 & P_{Load}^{elec} \leq C_{M,i,e} + C_{S,i,e} \end{cases} \tag{22}
$$

The energy supply expectation of power, heat, and cold energy types in the microgrid are calculated by Equations (23)–(25), respectively:

$$
EES^{elec} = \frac{1}{N} \sum_{j=1}^{N} Q_{j,i,e} \tag{23}
$$

$$
EES^{heat} = \frac{1}{N} \sum_{j=1}^{N} Q_{j,i,h} \tag{24}
$$

$$
EES^{cold} = \frac{1}{N} \sum_{j=1}^{N} Q_{j,i,c} \tag{25}
$$

where N is the number of Monte Carlo simulated sampling time periods.

(2) LOEE

This index denotes the gross loss of energy due to unit failure or outage of a certain energy source in the statistical time period. The unit is MW·h/a. A higher LOEE suggests that the energy source is greatly influenced by unit failure, and that energy supply reliability is low. The index is calculated by Equations (26)–(28).

$$LOEE^{\text{elec}} = \frac{8760}{T} \sum_{j=1}^{N} LOEE_i^{\text{elec}} \tag{26}$$

$$LOEE^{\text{heat}} = \frac{8760}{T} \sum_{j=1}^{N} LOEE_i^{\text{heat}} \tag{27}$$

$$LOEE^{\text{cold}} = \frac{8760}{T} \sum_{j=1}^{N} LOEE_i^{\text{cold}} \tag{28}$$

where T is the evaluation time cycle.

(3) SAIDI

SAIDI refers to the duration of insufficient supply of a certain type of energy caused by unit failure or outage in the statistical time period. The unit is h/a. Longer duration means a larger impact of unit failure and low-energy supply reliability. This index is calculated by Equations (29) and (31):

$$SAIDI^{\text{elec}} = \frac{8760}{T} \sum_{j=1}^{N} SAIDI_i^{\text{elec}} \tag{29}$$

$$SAIDI^{\text{heat}} = \frac{8760}{T} \sum_{j=1}^{N} SAIDI_i^{\text{heat}} \tag{30}$$

$$SAIDI^{\text{cold}} = \frac{8760}{T} \sum_{j=1}^{N} SAIDI_i^{\text{cold}} \tag{31}$$

(4) Energy supply reliability economic indicator

This study determined energy supply reliability economic indicators by calculating the LOEE by type with corresponding prices [27]. The index value reflects the economic loss expectation caused by equipment failure to the multi-energy microgrid system. It is calculated as follows:

$$C_{\text{RE}} = w^{\text{elec}} LOEE^{\text{elec}} + w^{\text{heat}} LOEE^{\text{heat}} + w^{\text{cold}} LOEE^{\text{cold}} \tag{32}$$

where w^{elec}, w^{heat}, and w^{cold} are the energy loss values of electricity, heat, and cold, respectively. The specific values can be extracted from the references [18].

4.3. FMEA Analysis for Reliability Evaluation

Due to space limitations, this study report here only analyses the impact of CCHP failure as well as quantitative analysis of the complementary coupling relationship between different energy types.

CCHP failure influences the supply of cold, heat, and power energy. Power loads are mainly supplied by the PV, with a backup supplement from ES devices and PG; cold loads are supplied by CAC under the condition that there is extra power energy and that CAC output constraints are satisfied. Heat loads are supplied by the GHP under the condition that the GHP output constraints are satisfied with backup support from HS devices.

Based on the evaluation indexes introduced in Section 4.2, the LOEE can be calculated by Equations (33) and (34):

$$LOEE_{\text{CCHP}}^{\text{elec}} = \sum_{k_{e,\text{CCHP}}} \left(\lambda_{\text{CCHP}} \cdot LOEE_{k_{e,\text{CCHP}}}^{\text{elec}} \right) \tag{33}$$

$$LOEE_{k_{e,CCHP}}^{\text{elec}} = \begin{cases} 0 & P_{\text{Load}}^{\text{elec}}(t) \leq P_{\text{output}}^{\text{ES}}(t) + P_{\text{output}}^{\text{PV}}(t) + P_{\text{output}}^{\text{EN}}(t) \\ \int_{r_{\text{CCHP}}} \begin{aligned} &[P_{\text{Load}}^{e}(t) - (P_{\text{output}}^{\text{ES}}(t)+ \\ &P_{\text{output}}^{\text{PV}}(t) + P_{\text{output}}^{\text{EN}}(t))]dt \end{aligned} & P_{\text{Load}}^{\text{elec}}(t) > P_{\text{output}}^{\text{ES}}(t) + P_{\text{output}}^{\text{PV}}(t) + P_{\text{output}}^{\text{EN}}(t) \end{cases} \tag{34}$$

SAIDI can be calculated by Equations (35) and (36):

$$SAIDI_{\text{CCHP}}^{\text{elec}} = \sum_{k_{e,CCHP}} (\lambda_{\text{CCHP}} \cdot SAIDI_{k_{e,CCHP}}^{\text{elec}}) \tag{35}$$

$$SAIDI_{k_{e,CCHP}}^{\text{elec}} = \begin{cases} 0 & P_{\text{Load}}^{\text{elec}}(t) \leq P_{\text{output}}^{\text{ES}}(t) + P_{\text{output}}^{\text{PV}}(t) + P_{\text{output}}^{\text{EN}}(t) \\ r_{\text{CCHP}} & P_{\text{Load}}^{\text{elec}}(t) > P_{\text{output}}^{\text{ES}}(t) + P_{\text{output}}^{\text{PV}}(t) + P_{\text{output}}^{\text{EN}}(t) \end{cases} \tag{36}$$

where $k_{e,CCHP}$ is the power supply area affected by the fault of the CCHP system.

Likewise, the relevant reliability indexes of heat and cold energy types can be calculated by Equations (37)–(40):

$$LOEE_{k_{h,CCHP}}^{\text{heat}} = \begin{cases} 0 & P_{\text{Load}}^{\text{heat}}(t) \leq P_{\text{output}}^{\text{GHP}}(t) + P_{\text{output}}^{\text{HS}}(t) \\ \int_{r_{\text{CCHP}}} \begin{aligned} &[P_{\text{Load}}^{\text{heat}}(t) - (P_{\text{output}}^{\text{GHP}}(t)+ \\ &P_{\text{output}}^{\text{HS}}(t))]dt \end{aligned} & P_{\text{Load}}^{\text{heat}}(t) > P_{\text{output}}^{\text{GHP}}(t) + P_{\text{output}}^{\text{HS}}(t) \end{cases} \tag{37}$$

$$LOEE_{k_{c,CCHP}}^{\text{cold}} = \begin{cases} 0 & \left\{P_{\text{Load}}^{\text{elec}}(t) < P_{\text{output}}^{\text{ES}}(t) + P_{\text{output}}^{\text{PV}}(t) + P_{\text{output}}^{\text{EN}}(t)\right\} \& \left\{P_{\text{Load}}^{\text{cold}}(t) < P_{\text{output}}^{\text{CAC}}(t)\right\} \\ \int_{r_{\text{CCHP}}} [P_{\text{Load}}^{\text{cold}}(t) - P_{\text{output}}^{\text{CAC}}(t)]dt & \left\{P_{\text{Load}}^{\text{elec}}(t) < P_{\text{output}}^{\text{ES}}(t) + P_{\text{output}}^{\text{PV}}(t) + P_{\text{output}}^{\text{EN}}(t)\right\} \& \left\{P_{\text{Load}}^{\text{cold}}(t) \geq P_{\text{output}}^{\text{CAC}}(t)\right\} \\ \int_{r_{\text{CCHP}}} [P_{\text{Load}}^{\text{cold}}(t)]dt & P_{\text{Load}}^{\text{elec}}(t) \geq P_{\text{output}}^{\text{ES}}(t) + P_{\text{output}}^{\text{PV}}(t) + P_{\text{output}}^{\text{EN}}(t) \end{cases} \tag{38}$$

$$SAIDI_{k_{h,CCHP}}^{\text{heat}} = \begin{cases} 0 & P_{\text{Load}}^{\text{heat}}(t) \leq P_{\text{output}}^{\text{GHP}}(t) + P_{\text{output}}^{\text{HS}}(t) \\ r_{\text{CCHP}} & P_{\text{Load}}^{\text{heat}}(t) > P_{\text{output}}^{\text{GHP}}(t) + P_{\text{output}}^{\text{HS}}(t) \end{cases} \tag{39}$$

$$SAIDI_{k_{c,CCHP}}^{\text{cold}} = \begin{cases} 0 & \left\{P_{\text{Load}}^{\text{elec}}(t) < P_{\text{output}}^{\text{ES}}(t) + P_{\text{output}}^{\text{PV}}(t) + P_{\text{output}}^{\text{EN}}(t)\right\} \& \left\{P_{\text{Load}}^{\text{cold}}(t) < P_{\text{output}}^{\text{CAC}}(t)\right\} \\ & \left\{P_{\text{Load}}^{\text{elec}}(t) < P_{\text{output}}^{\text{ES}}(t) + P_{\text{output}}^{\text{PV}}(t) + P_{\text{output}}^{\text{EN}}(t)\right\} \& \left\{P_{\text{Load}}^{\text{cold}}(t) \geq P_{\text{output}}^{\text{CAC}}(t)\right\} \\ r_{\text{CCHP}} & \cup \\ & P_{\text{Load}}^{\text{elec}}(t) \geq P_{\text{output}}^{\text{ES}}(t) + P_{\text{output}}^{\text{PV}}(t) + P_{\text{output}}^{\text{EN}}(t) \end{cases} \tag{40}$$

According to the cooling energy determined by Equations (38) and (40), a portion of the cooling loads are supplied by the CAC during normal function. If there is a fault, the system, in terms of energy loads of equal importance, gives priority to electric loads per the higher energy grade of electric power than cooling energy, besides their rigidity and flexibility. In contrast, the cooling loads supplied by the CAC are reduced when a malfunction occurs to ensure the continuous supply of electric loads.

Likewise, the impacts of other unit failures on different energy supplies in the multi-energy microgrid can be analyzed. The effects of PG failures on the microgrid system as well as the supply $P_{\text{output}}^{\text{PG}}$ in the electric-isolated operation mode are not discussed here.

5. Multi-Energy Microgrid Double-Layer Planning Model

The optimized planning model of the multi-energy microgrid has two major parts. First is the optimization at the planning level, which ensures the favorable capacity and quality of energy supply and storage devices in the microgrid system. Second is the optimization at the scheduling level, which serves to optimize the output of power supply devices and the operation of energy storage devices in the microgrid system. Both parts target economical efficiency per the reliability factors of the power supply.

5.1. Optimization Object

In this study, the total cost of the multi-energy microgrid system across its entire life cycle was selected as the target function. The total cost primarily involves the purchase, installation cost, operation cost, and reliability conversion cost. The target function is expressed as:

$$\min\{C_{\text{IN}} + C_{\text{OP}} + C_{\text{RE}}\} \tag{41}$$

where C_{IN} is the installation cost; C_{OP} is the operation cost; and C_{RE} is the reliability conversion cost. Where the system construction cost is:

$$C_{\text{IN}} = \left(\sum_m N_m \times C_{inst,m} \times M_m - \sum_m N_m \times C_{scra,m} \right) \frac{r(1+r)^Y}{(1+r)^Y - 1} \tag{42}$$

where $C_{inst,i}$ is the investment and construction cost of a unit capacity of mth equipment unit; m is the total number of equipment units; and r is the effective rate of interest, which is set as 5%.

The annual operation cost is:

$$C_{\text{OP}} = \sum_m \sum_t C_{fuel,m}^{OP}(t) + C_{opma,m}^{OP}(t) \tag{43}$$

where $C_{\text{opma},m}^{\text{OP}}(t)$ is the operating and maintenance cost of mth equipment unit at time interval t.

The calculation method of the reliability conversion cost C_{RE} is expressed as Equation (32).

5.2. Constraints

(1) Maximum load constraint: The maximum output capacity of the energy supply equipment configuration must meet the maximum power demand of the microgrid system in power, heating, and cooling loads:

$$\sum_m P_{\text{output,max}}^m(t) \geq P_{\text{Load,max}}^{[\cdot]}(t) \tag{44}$$

where $P_{\text{output,max}}^m(t)$ is the maximum output power of mth energy supply equipment unit at time interval t; and $P_{\text{Load,max}}^{[\cdot]}(t)$ is the maximum load demand for some kind of energy at time interval t.

(2) Power balancing constraint: If the energy storage function is ensured, a real-time balance of power/heating/cooling energy supply and demand in the multi-energy microgrid should be achieved:

$$\sum_m P_{\text{output}}^m(t) + \sum_m P_{\text{D}}^m(t) = P_{\text{Load}}^{[\cdot]}(t) + \sum_m P_{\text{C}}^m(t) \tag{45}$$

where $P_{\text{output}}^{[\cdot]}(t)$ is the output power of mth energy supply equipment unit at time interval t; $P_{[\text{C/D}]}^m(t)$ is the charge or discharge power of mth energy storage equipment unit at time interval t; and $P_{\text{Load}}^{[\cdot]}(t)$ is the load demand at time interval t.

(3) Self-sufficiency probability constraint: The multi-energy microgrid works in electric-isolated mode, so it is crucial in regard to meeting the load demands and operation stability of the system. The self-sufficiency probability of meeting the system load demand in the planning period can be employed to direct the planning and configuration of the microgrid system:

$$\Pr\left(\sum_m P_{\text{output}}^m(t) + \sum_m P_{\text{D}}^m(t) \geq P_{\text{Load}}^{[\cdot]}(t) \right) \geq P_{\text{st}}^{[\cdot]} \tag{46}$$

where $P_{\text{st}}^{[\cdot]}$ is the probability of self-sufficiency of a certain type of energy load.

(4) Operation constraint of power supply equipment: The energy supply equipment of the multi-energy microgrid must meet the rated power and climbing constraints of the equipment during operation.

$$P_{\text{output,min}}^m \leq P_{\text{output}}^m(t) \leq P_{\text{output,max}}^m \tag{47}$$

$$-\Delta P_{\text{down}}^m \leq P_{\text{output}}^m(t) - P_{\text{output}}^m(t-1) \leq \Delta P_{\text{up}}^m \tag{48}$$

where $-\Delta P_{\text{down}}^m$ and ΔP_{up}^m are the climbing power of the mth energy supply component to reduce or increase output power.

(5) Operation constraint of power storage equipment: The energy storage equipment of the multi-energy microgrid must meet the energy charge/discharge power, and capacity constraints of the equipment during operation:

$$M_{\text{min}}^m(t) \leq M^m(t) \leq M_{\text{max}}^m(t) \tag{49}$$

$$P_{[\text{C/D}]}^m(t) \leq P_{[\text{C/D}],\text{max}}^m(t) \tag{50}$$

where $M_{\text{max}}^m(t)$ and $M_{\text{min}}^m(t)$ are the maximum or minimum capacity of the mth energy storage equipment unit.

5.3. Multi-Energy Microgrid Planning Process

The planning model of the multi-energy microgrid optimization also has two parts: planning optimization and operation scheduling optimization. The configuration scheme is optimized based on the overall energy load level in the microgrid system under the given constraints. In alternative energy supply and storage equipment, considering Equation (42) and the optimal operation conditions in the system's whole life cycle, an optimized combination scheme for energy supply devices and energy storage devices of the microgrid system is proposed here with the equipment type and quantity as optimization variables. The planning optimization denotes a non-linear integer programming algorithm that does not readily provide an analytical solution, so this study instead employed the quantum-behaved particle swarm optimization (QPSO) algorithm [28]. Considering that the velocity vectors are usually assumed to be small values during the iterative process of the traditional PSO algorithm, one of the main drawbacks of the traditional PSO algorithm is that the solution of the problem is easy to trap into the local optimum rather than the global optimal. Therefore, in this paper, the QPSO algorithm is employed to avoid this problem. By using the wave function to describe the state of a particle instead of the velocity as in a traditional PSO algorithm, and by improving the dynamic behavior of particles, the solution of the problem can effectively avoid falling into local optimum.

Next, based on the hourly load demand curve in the planning period, the power supply of the microgrid system, the demand profile of the microgrid system, and the operation constraints in the energy supply/storage equipment, an hourly optimization scheduling scheme for each component is built by combining the equipment configuration scheme given in the planning part and taking the start/stop and output of the energy supply and storage equipment. Scheduling optimization denotes a mixed non-linear optimization problem, which can also be solved through the QPSO algorithm. Economical indicators under a certain planning scheme throughout the system's operation can be obtained at the operation level. Thus, they play an important part in the economical target of microgrid system planning. Optimizing the operation and scheduling level is virtually a sub-planning-level optimization process.

The optimized operation and scheduling level denotes the economic performance of the microgrid system under normal operation conditions. The economical efficiency converted by the energy supply reliability of the microgrid system is also calculated to determine the economic loss expectation produced under fault conditions and improve the economical indexes. This process is illustrated in Figure 2.

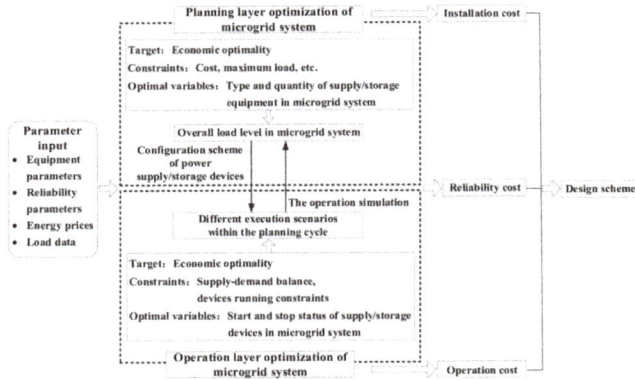

Figure 2. Optimal planning process of a multi-energy microgrid.

6. Case Study

6.1. Case Overview

This study employed a series of typical industrial parks in southern China to validate the proposed method. The physical structure and equipment composition of the microgrid system is given in Figure 1. In regard to energy supply and demand, April to October are classified as cooling months (i.e., with large demand for cooling loads). Heating loads include the demand for drying and ventilation during the production and hot water for residential life; they have no definite supply period, but still show some seasonal characteristics. There are demands for electric loads throughout the year.

Curves reflecting the demands for electric, heating, and cooling loads in the microgrid system throughout the year are shown in Figure 3. The available types of energy production/conversion equipment and economical operating parameters in the microgrid based on the above demands are listed in Table 1; the types and economical operating parameters of energy storage devices are listed in Table 2. The initial capacities of ES and HS are 30% and 50% of the rated capacity, respectively; the maximal charge–discharge power is 80% of the rated capacity [29]. The reliability parameters of various equipment units are listed in Table 3.

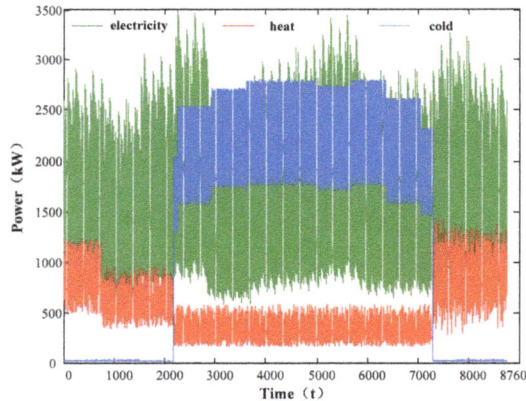

Figure 3. Load demand curve.

Table 1. Types and parameters of alternative energy production/conversion equipment. CCHP: combined cold heat and power, GHP: gas heat pump, CAC: central air conditioning.

Devices Types	Capacity/kW	Initial Investment Cost/(CNY·kW^{-1})	Operation and Maintenance Cost/(CNY·kW^{-1})	Energy Conversion Efficiency	Upper Limit/kW	Lower Limit/kW
CCHP 1	1200	4780	1.558	1	1200	12
CCHP 2	1500	4700	1.554	1	1500	15
CCHP 3	2000	4590	1.547	1	2000	20
CCHP 4	3000	3500	1.499	1	3000	30
GHP 1	800	2430	0.40	80%	800	8
GHP 2	1000	2230	0.40	80%	1000	10
GHP 3	1500	1680	0.40	80%	1500	15
CAC 1	1400	511	0.30	88%	1400	14
CAC 2	2000	460	0.30	90%	2000	20

Table 2. Type and parameters of energy storage equipment. ES: electric storage, HS: heat storage.

Device Type	Capacity/kW	Initial Investment Cost/(CNY·kW^{-1})	Operation and Maintenance Cost/(CNY·kW^{-1})	Initial Capacity/kW·h	Maximum Power/kW
ES 1	1000	700	0.03	300	800
ES 2	3000	580	0.03	1000	2400
HS	1000	450	0.03	500	800

Table 3. Reliability parameters. PV: photovoltaic, GN: gas network.

Devices Names	λ_k (f/a)	r_k (h)
CCHP	4	24
GHP	0.6	2
CAC	0.4	2
PV	0.4	20
ES	0.05	50
GN	0.9	20

The gas network sets the fault rate and repair time of the main gas pipeline. Various factors correlated with power-generating efficiency are discussed here. The installed capacity of the PV is 4.6 MW; the annual output curve is shown in Figure 4. The optimization of different PV configurations is not discussed in subsequent planning programs. According to the local power pricing policy, the power price peaks at 11:00–15:00 and 19:00–21:00, and then valleys at 0:00–7:00; in other periods, the price remains stable. Table 4 lists the prices for different energy types. The loss value of electric/heating/cooling loads in the microgrid are 200 ¥/kW·h, 120 ¥/kW·h, and 120 ¥/kW·h, respectively. Self-sufficiency probability is 90%.

Table 4. Different energy prices.

Types of Energy		Energy Prices CNY/kW·h; CNY/m^3
electricity	Peak period	0.80
	Waist period	0.50
	Valley period	0.20
gas		2.28

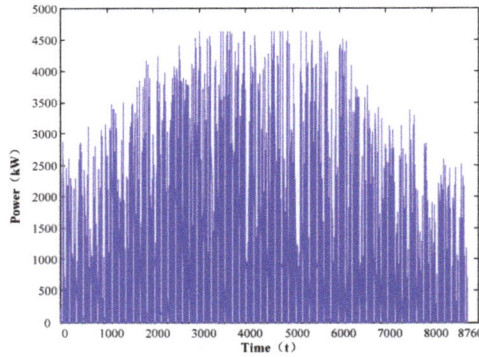

Figure 4. Photovoltaic (PV) output curve.

6.2. Configuration of Storage Devices

The microgrid system described in this study operates in electric-isolated mode and in a self-sufficient manner. The overall time cycle of the scheme is planned to be 10 years, i.e., $Y = 10a$. The 10th year is considered the target year. It is also assumed that installation occurs in the first year of the planning horizon [30]. The penalty coefficient of the relevant algorithms is 0.03, the population scale is 80, the inertia factor is 0.5, the self factor is 2, the global factor is 2, the mutation probability is 0.05, the iteration number N is 1000, and the error precision σ is 0.01. Four different cases were employed to analyze the impacts of different configurations on the selection and reliability of different microgrid planning schemes.

Case 0: A microgrid system without ES or HS devices;
Case 1: A microgrid system with HS devices but lacking ES devices;
Case 2: A microgrid system with ES devices but lacking HS devices;
Case 3: A microgrid equipped with both ES and HS devices.

This study calculated the optimal configuration scheme according to the above four cases; see Tables 5 and 6 for their respective results and costs. Case 3 has the best economical performance considering the capacity, configuration, operation, and reliability of the microgrid system. Case 0 has low investment costs, as it lacks energy storage devices. In this case, the energy supply equipment of the microgrid system is forced to meet the load demands at any moment and deal with the intermittency of distributed PV power, resulting in high operation costs. The absence of energy storage devices also leads to low reliability, especially when malfunction occurs, which increases the loss expectation.

Table 5. Optimal configuration of microgrid system.

Case	CCHP	GHP	CAC	ES	HS
0	CCHP 4	GHP 3	CAC 2	-	-
1	CCHP 4	GHP 2	CAC 2	-	HS
2	CCHP 3	GHP 2	CAC 2	ES 2	-
3	CCHP 3	GHP 2	CAC 2	ES 2	HS

Table 6. Cost calculation result. Unit: CNY.

Case	Total Cost	C_{IN}	C_{OP}	C_{RE}
0	8.103×10^6	1.805×10^6	4.093×10^6	2.204×10^6
1	7.165×10^6	1.826×10^6	4.126×10^6	1.213×10^6
2	6.464×10^6	1.824×10^6	3.9×10^6	0.757×10^6
3	6.028×10^6	1.880×10^6	3.451×10^6	0.696×10^6

Cases 1 and 2 have different energy storage devices. The HS device is a pure backup resource that improves heating reliability. The ES device can stabilize the volatility of PV, which can improve the microgrid system's power supply reliability and further enhance the reliability of other energy resources by using the energy conversion devices. Accordingly, the ES device is more important than the HS device.

6.3. Reliability Assessment

Reliability is an important indicator for any ideal planning configuration scheme. Therefore, this study carefully assessed the reliability of the energy supply to the system based on the above optimized configuration scheme. The energy supply reliability of the microgrid system under different cases were compared per the reliability indicators discussed in Section 4.2. The calculation results are shown in Table 7 and Figures 5 and 6.

Table 7. Comparison of expectation of energy supplied (EES) in different cases.

Case	Electricity	Heat	Cold
0	0.9956	0.9953	0.9937
1	0.9977	0.9979	0.9963
2	0.9985	0.9974	0.9972
3	0.9987	0.9982	0.9977

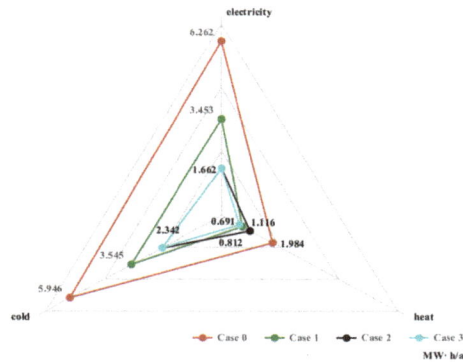

Figure 5. Comparison of loss of energy expectation (LOEE) in different scenes.

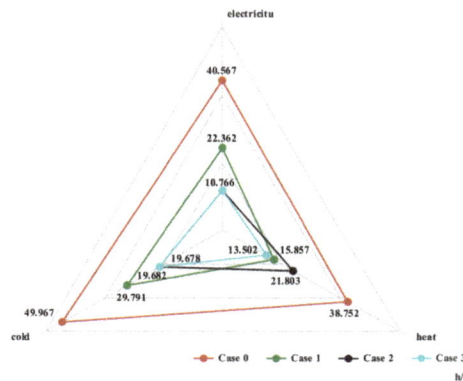

Figure 6. Comparison of system average interruption duration index (SAIDI) in different cases.

In Case 1, the system lacks the backup of energy storage devices and meets energy demands through the cooperation of relevant equipment units within the microgrid system. The distributed PV power with high volatility cannot be stabilized effectively. Accordingly, energy supply reliability is poor under random fault due to the lower degree of coupling among energy sources. A reasonable configuration of energy storage devices in the microgrid can effectively stabilize the output randomness of renewable energy sources, reduce the capacity redundancy of the energy production devices, improve the energy supply reliability, and ensure an economical system performance. The HS device is a backup resource that is critical for heat supply reliability. The ES device not only improves the reliability of the power supply, it also converts energy sources through CAC when necessary. Power storage devices improve the coupling among energy sources in the microgrid system and indirectly enhance the energy supply reliability of other energy sources in the system.

Typical optimal configuration schemes under Case 4 were selected for further analysis in accordance with their favorable annual reliability. Preconceived CCHP faults were simulated to explore the impacts on all of the power, heating, and cooling resources with multi-energy coupling on energy supply reliability under different energy demands. This study selected the peak and valley periods of power loads, heat loads, and cold loads in a typical year, simulated CCHP outage and failure, and then evaluated the LOEE and outage impact time expectation (r_e) of the multi-energy microgrid under different energy demands. A pre-arranged outage time according to the reference value of failure repair, 24 h, was employed to ensure that the expected failure time did not affect the results. The evaluations are listed in Table 8, where e-peak, h-peak, and c-peak denote the peak periods of power loads, heat loads, and cold loads, respectively; likewise, the valley periods of loads are represented.

Table 8. Reliability indexes of a typical period of time for the microgrid system. LOEE: loss of energy expectation, r_e: outage impact time expectation.

Reliability Indexes		e-Peak	e-Valley	h-Peak	h-Valley	c-Peak	c-Valley
LOEE (MW·h)	electricity	1.985	1.7942	1.434	1.187	1.883	2.482
	heat	0	0	0.129	0	0	0.959
	cold	3.847	5.138	1.629	3.995	5.009	0.047
r_e (h)	electricity	18	21	12	12	16	18
	heat	0	0	5	0	0	10
	cold	21	24	18	23	24	21

When performing pre-arranged outage maintenance of a traditional single energy network, the goal is typically to avoid load peaks. However, the comparison of reliability indexes at different typical periods under the same operation mode suggests that although the pre-arranged valley outage value does partially ensure sufficient energy supply, it significantly impacts other energy supplies under multi-energy coupling and energy supply cycle circumstances. The valley value of power loads may actually have a greater impact on cold loads than the simulated peak failure value.

On this basis, the energy supply reliability economic indicator was further analyzed to determine the optimal maintenance time for major equipment in the microgrid system. Through the calculation, the preconceived accident simulation is conducted at the moment $t = 3500$, and can get $I_{min} = 1332.631$ CNY. On account of the high PV output in the simulation period, ES can effectively reduce the loss of energy supply. Extra power can be employed for cold loads through CAC, and heat loads can be transferred completely through GHP. Hence, the influence on the microgrid system at this moment is lower than that of other moments.

6.4. Discussions

In the case study, this study quantitatively analyzed the influence of different equipment configuration schemes on the economy and reliability of multi-energy microgrid planning from different perspectives. On the basis, the reliability of a pre-scheduled outage in a typical scene was analyzed, and a kind of application scenario of reliability evaluation was put forward. The results of the cases are discussed as follows:

(1) According to the analysis of Section 6.2, although energy storage equipment will increase the installation cost of the microgrid system to some extent, energy storage equipment can stabilize the volatility of the renewable energy output and optimize energy distribution. Therefore, the operation cost of the microgrid system is improved, and the energy storage equipment can play the role of a backup resource, which enhances the reliability of the microgrid system. Considering the reliability conversion cost, the configuration of the energy storage equipment can improve the total cost of the microgrid system.

(2) The analysis of Section 6.3 suggests that compared with ES, HS equipment can only play the role of heat energy reserve resources, so the contribution to reliability promotion is limited. The ES equipment can not only provide support for improving the reliability of a power supply, it can also indirectly enhance the reliability of the cooling supply because of the coupling characteristics of the multi-energy networks through the equipment.

(3) Based on the analysis of the reliability of the multi-energy microgrid system, the reliability within a pre-arranged outage in a typical scene was further analyzed in Section 6.3, which can be referenced by the establishment of a maintenance strategy for key equipment in a microgrid system.

7. Conclusions

This study took a multi-energy microgrid as its research target, sorted out the energy efficiency and economical models of the key equipment in the microgrid system, presented a double-layer model with optimal configuration, operation, and reliability consideration for the multi-energy microgrid planning, and determined the optimal capacity configuration and energy management scheme via an optimization algorithm. Compared with traditional, independently supplied and operated energy supply methods, joint planning and design considering multi-energy coupling and complement significantly improved the economic performance of the microgrid system. The energy storage device in the microgrid system mitigated fluctuations in the output of distributed power and further enhanced the energy supply reliability of the microgrid.

In the meantime, this study also proposed a calculation method for energy supply reliability under multi-energy circumstances and perfected the planning model in terms of the economic loss caused by reliability evaluations. The study properly calculated the significance of reliability in planning and compensated for drawbacks in the traditional planning model (which only considers economical factors under normal operation). The optimized design model with the noted reliability consideration can help decision-makers determine the optimal equipment capacity and quantity to satisfy the load demands in actual multi-energy microgrids. The proposed model was designed to not only improve energy supply reliability, but also minimize cost. On this basis, the paper further studied the function of reliability in the maintenance of key units. Based on the proposed reliability assessment techniques, the paper made a comprehensive analysis on the planning, operation, and maintenance of a multi-energy microgrid, providing certain guidance for the future construction of integrated energy systems.

For future research, regarding reliability evaluation, the impact of the system structure such as lines, pipelines on reliability, as well as the time-delay difference of different energy transmission will be further considered. Moreover, the impact of different operation modes of a microgrid system on reliability will be analyzed as well. In the research on planning methods, energy efficiency, pollutant emissions, and other factors will be further considered to improve the planning model from different perspectives.

Author Contributions: Conceptualization, S.G., J.L. and H.L.; methodology, J.L. and H.L.; J.L. and H.S.; validation, S.G. and H.L.; formal analysis, J.L.; writing—J.L. and Y.W.; writing—review and editing, S.G. and H.L.

Funding: This research was funded by National Natural Science Foundation of China, grant number 51777133.

Acknowledgments: This work was supported by the National Key R&D Plan of China (2017YFB0903400).

Conflicts of Interest: The author's team has confirmed that there is no conflict of interest in this paper.

References

1. Lin, W.; Jin, X.; Mu, Y.; Jia, H.; Xu, X.; Yu, X.; Zhao, B. A two-stage multi-objective scheduling method for integrated community energy system. *Appl. Energy* **2018**, *216*, 428–441. [CrossRef]
2. Geidl, M.; Koeppel, G.; Favre-Perrod, P.; Klöckl, B.; Andersson, G.; Fröhlich, K. Energy hubs for the future. *IEEE Power Energy Mag.* **2007**, *5*, 24–30. [CrossRef]
3. Wu, J.; Yan, J.; Jia, H.; Hatziargyriou, N.; Djilali, N.; Sun, H.B. Integrated Energy Systems. *Appl. Energy* **2016**, *167*, 155–157. [CrossRef]
4. Razeghi, G.; Gu, F.; Neal, R.; Samuelsen, S. A generic microgrid controller: Concept, testing, and insights. *Appl. Energy* **2018**, *229*, 660–671. [CrossRef]
5. Liu, T. Energy management of cooperative microgrids: A distributed optimization approach. *Int. J. Electr. Power Energy Syst.* **2018**, *96*, 335–346. [CrossRef]
6. Li, H.; Eseye, A.T.; Zhang, J.; Zheng, D. Optimal energy management for industrial microgrids with high-penetration renewables. *Prot. Control Mod. Power Syst.* **2017**, *2*, 1–12. [CrossRef]
7. Escalera, A.; Hayes, B.; Prodanović, M. A survey of reliability assessment techniques for modern distribution networks. *Renew. Sustain. Energy Rev.* **2018**, *91*, 344–357. [CrossRef]
8. Billinton, R.; Wang, P. Teaching distribution system reliability evaluation using Monte Carlo simulation. *IEEE Trans. Power Syst.* **1999**, *14*, 397–403. [CrossRef]
9. Che, L.; Zhang, X.; Shahidehpour, M.; Alabdulwahab, A.; Abusorrah, A. Optimal Interconnection Planning of Community Microgrids with Renewable Energy Sources. *IEEE Trans. Smart Grid* **2017**, *8*, 1054–1063. [CrossRef]
10. Xu, X.; Mitra, J.; Wang, T.; Mu, L. An Evaluation Strategy for Microgrid Reliability Considering the Effects of Protection System. *IEEE Trans. Power Deliv.* **2016**, *31*, 1989–1997. [CrossRef]
11. Chen, C.; Wu, W.; Zhang, B.; Singh, C. An Analytical Adequacy Evaluation Method for Distribution Networks Considering Protection Strategies and Distributed Generators. *IEEE Trans. Power Deliv.* **2015**, *30*, 1392–1400. [CrossRef]
12. Moeini-Aghtaie, M.; Farzin, H.; Fotuhi-Firuzabad, M.; Amrollahi, R. Generalized Analytical Approach to Assess Reliability of Renewable-Based Energy Hubs. *IEEE Trans. Power Syst.* **2016**, *32*, 368–377. [CrossRef]
13. Li, G.; Bie, Z.; Kou, Y.; Jiang, J.; Bettinelli, M. Reliability evaluation of integrated energy systems based on smart agent communication. *Appl. Energy* **2016**, *167*, 397–406. [CrossRef]
14. Koeppel, G.; Andersson, G. Reliability modeling of multi-carrier energy systems. *Energy* **2009**, *34*, 235–244. [CrossRef]
15. Koeppel, G.; Andersson, G. The influence of combined power, gas, and thermal networks on the reliability of supply. In Proceedings of the Sixth World Energy System Conference, Torino, Italy, 10–12 July 2006; pp. 646–651.
16. Chaudry, M.; Wu, J.; Jenkins, N. A sequential Monte Carlo model of the combined GB gas and electricity network. *Energy Policy* **2013**, *62*, 473–483. [CrossRef]
17. Adefarati, T.; Bansal, R.C. Reliability and economic assessment of a microgrid power system with the integration of renewable energy resources. *Appl. Energy* **2017**, *206*, 911–933. [CrossRef]
18. Shahmohammadi, A.; Moradi-Dalvand, M.; Ghasemi, H.; Ghazizadeh, M.S. Optimal Design of Multicarrier Energy Systems Considering Reliability Constraints. *IEEE Trans. Power Deliv.* **2015**, *30*, 878–886. [CrossRef]
19. Al Ghaithi, H.M.; Fotis, G.P.; Vita, V. Techno-economic assessment of hybrid energy off-grid system—A case study for Masirah island in Oman. *Int. J. Power Energy Res.* **2017**, *1*, 103–116. [CrossRef]
20. Rana, M.M.; Xiang, W.; Wang, E. Smart grid state estimation and stabilisation. *Int. J. Electr. Power Energy Syst.* **2018**, *102*, 152–159. [CrossRef]
21. Zhang, X.; Shahidehpour, M.; Alabdulwahab, A.; Abusorrah, A.M. Optimal Expansion Planning of Energy Hub with Multiple Energy Infrastructures. *IEEE Trans. Smart Grid* **2015**, *6*, 2302–2311. [CrossRef]

22. Conti, S.; Nicolosi, R.; Rizzo, S.A. Generalized Systematic Approach to Assess Distribution System Reliability with Renewable Distributed Generators and Microgrids. *IEEE Trans. Power Deliv.* **2012**, *27*, 261–270. [CrossRef]

23. Zhao, B. *Key Technology and Application of Microgrid Optimal Configuration*; Science Press: Beijing, China, 2015.

24. Dou, C.X.; Liu, B. Multi-Agent Based Hierarchical Hybrid Control for Smart Microgrid. *IEEE Trans. Smart Grid* **2013**, *4*, 771–778. [CrossRef]

25. Korkas, C.D.; Baldi, S.; Michailidis, I.; Kosmatopoulos, E.B. Occupancy-based demand response and thermal comfort optimization in microgrids with renewable energy sources and energy storage. *Appl. Energy* **2016**, *163*, 93–104. [CrossRef]

26. Zhong, W.; Xie, K.; Liu, Y.; Yang, C. Auction Mechanisms for Energy Trading in Multi-Energy Systems. *IEEE Trans. Ind. Inform.* **2017**, *14*, 1511–1521. [CrossRef]

27. Xu, X.; Hou, K.; Jia, H.; Yu, X. A reliability assessment approach for the urban energy system and its application in energy hub planning. In Proceedings of the IEEE Power & Energy Society General Meeting, Denver, CO, USA, 26–30 July 2015.

28. Mohanty, C.P.; Mahapatra, S.S.; Singh, M.R. An intelligent approach to optimize the EDM process parameters using utility concept and QPSO algorithm. *Eng. Sci. Technol. Int. J.* **2017**, *20*, 552–562. [CrossRef]

29. Erdinc, O. Economic impacts of small-scale own generating and storage units, and electric vehicles under different demand response strategies for smart households. *Appl. Energy* **2014**, *126*, 142–150. [CrossRef]

30. Farret, F.A.; Simões, M.G. *Integration of Alternative Sources of Energy*; Wiley: Hoboken, NJ, USA, 2006.

![applied sciences logo] *applied sciences*

MDPI

Article

Optimal Operation of Isolated Microgrids Considering Frequency Constraints

Josep-Andreu Vidal-Clos, Eduard Bullich-Massagué *, Mònica Aragüés-Peñalba,
Guillem Vinyals-Canal, Cristian Chillón-Antón, Eduardo Prieto-Araujo and
Oriol Gomis-Bellmunt and Samuel Galceran-Arellano

Centre d'Innovació Tecnològica en Convertidors Estàtics i Accionaments (CITCEA-UPC),
Departament d'Enginyeria Elèctrica, Universitat Politècnica de Catalunya ETS d'Enginyeria Industrial de
Barcelona, Avinguda Diagonal, 647, Pl. 2, 08028 Barcelona, Spain; josep.andreu.vidal@citcea.upc.edu (J.-A.V.-C.);
monica.aragues@citcea.upc.edu (M.A.-P.); guillem.vinals.canal@citcea.upc.edu (G.V.-C.);
cristian.chillon@citcea.upc.edu (C.C.-A.); eduardo.prieto-araujo@citcea.upc.edu (E.P.-A.);
gomis@citcea.upc.edu (O.G.-B.); galceran@citcea.upc.edu (S.G.-A.)
* Correspondence: eduard.bullich@citcea.upc.edu; Tel.: +34-93-405-4245

Received: 8 December 2018; Accepted: 3 January 2019; Published: 9 January 2019

Abstract: Isolated microgrids must be able to perform autonomous operation without external grid support. This leads to a challenge when non-dispatchable generators are installed because power imbalances can produce frequency excursions compromising the system operation. This paper addresses the optimal operation of PV–battery–diesel-based microgrids taking into account the frequency constraints. Particularly, a new stochastic optimization method to maximize the PV generation while ensuring the grid frequency limits is proposed. The optimization problem was formulated including a minimum frequency constraint, which was obtained from a dynamic study considering maximum load and photovoltaic power variations. Once the optimization problem was formulated, three complete days were simulated to verify the proper behavior. Finally, the system was validated in a laboratory-scaled microgrid.

Keywords: energy management system; microgrids; frequency stability; renewable power generation

---

1. Introduction

The integration of distributed generation requires the development of new concepts for active grid operation, where microgrids are the most promising one [1]. Microgrids can operate in grid connected as well as isolated modes [2,3]. In isolated mode, the active power balance to maintain the grid frequency has become one of the main challenges. The integration of large amount of photovoltaic (PV) generation can further stress the power balance due to the lack of inertia and the fast power variations of the resource. One possible solution to avoid frequency deviations produced by PV power generation is its curtailment [4]. Frequency deviations can also be limited by increasing the grid inertia, which can be achieved by connecting rotating machines [5]. The main drawback is that these solutions have and adverse effect on the operation cost.

To solve the power balance problems while minimizing the operation cost, a hierarchical control architecture is commonly used [6–9]. The primary control layer stabilizes the voltage and frequency deviations due to power imbalances by adjusting the active and reactive power references in a time frame of milliseconds. Then, the secondary control is responsible for recovering the voltage and frequency to their reference values. Commonly, it is done by using PI based closed loop controllers in a slower time scale than the primary control response time. Finally, The tertiary control determines the power references to perform the optimal operation of the microgrid.

1.1. Literature Review

Different methods have been considered for designing energy management systems (EMS), i.e., the tertiary control layer, for microgrids. These methods mainly consist of: (i) formulating an objective function; (ii) defining a set of constraints to ensure the proper system behavior; and (iii) applying an algorithm to find the optimal solution.

In [10], a mixed integer linear program (MILP) is formulated to minimize the microgrid operation cost. The microgrid includes critical and controllable loads, energy storage, controllable generation and renewable generation. Because the system under study is connected to the utility grid, any power imbalance is considered to be compensated by the external network producing a very small frequency deviation. Accordingly, power reserves are not considered. Even though the problem formulation does not consider forecast errors, its periodical execution, similar to the rolling horizon process, permits redefining periodically the operation plan compensating unpredicted deviations.

The works presented in [11–13] prove the real implementation of different EMSs for the minimum price or minimum cost of the isolated microgrid operation. These papers solve the optimization problems using MILP, multi-layer ant colony optimization and multi-period gravitational search algorithms, respectively. These studies consider perfect forecast. Thus, the hierarchical control structure is not implemented and power reserves are not considered. Consequently, power imbalances and frequency deviations are not studied. Thus, the grid stability cannot be ensured.

The study performed in [14] proposes a heuristic method, based on genetic algorithms, for solving the cost minimization problem for the microgrid operation. It first develops a forecasting method and then formulates the problem and the generic algorithm. The problem formulation differs depending on whether the microgrid operates connected or disconected form the main grid, considering load and generation forecast for the power balance equations. As power reserves are considered, the power imbalances due to forecast errors may be compensated, but this may lead to a suboptimal operation point of the microgrid. In addition, the transient response when imbalances due to forecast errors occur is not analyzed.

To avoid operating in a suboptimal operation point in microgrids due to forecast errors, different studies propose the formulation of stochastic optimization problems [15–17]. In this method, a set of forecasted scenarios is generated. Then, the decision variables are optimized for all scenarios, where the objective function is the sum of the objective function of each scenario.

In [18], an EMS for minimizing the use of diesel generation in a PV–wind–diesel–battery-based isolated microgrid is developed. The optimization problem is formulated as a MILP and executed using the rolling horizon technique to reduce the effects of the uncertainties of forecasted variables. In addition, the primary control layer (particularly the droop curves) vary depending if diesel generation is turned on or off. This fact can affect the transient performance, but a transient study is not performed.

The authors of [11–13] did not consider forecast errors. This issue is solved in [14,18] by considering power reserves. To improve the average optimal operation point against the uncertainty, the authors of [15–17] proposed a stochastic optimization method. These previous studies do not analyze dynamic and transient behavior. This gap is treated in [19]. This study develops a multi agent EMS for an isolated microgrid. One of the particularities and not studied in the previous cited papers, is that the transient response considering the primary and secondary control layers is analyzed. The tertiary control layer (EMS), which is the objective of the study, determines not only the scheduled setpoints but also the required reserves to compensate photovoltaic and load forecasting errors, avoiding frequency deviations. These frequency deviations are analyzed later in a real time dynamic simulator platform.

The local controls of generation units will react to frequency deviations to achieve a power balance and to maintain the grid frequency. In [20], the system frequency is introduced into the optimization problem. Particularly, the frequency-power (f-P) droop control is considered and a the maximum frequency deviation is constrained. These constraints apply for the steady state, but they do not consider the transient behavior. An optimal power flow (OPF) problem that includes the frequency transient behavior is presented in [21], explaining the need to limit its deviations. However, the main assumption is that the frequency decreases linearly during the first few seconds until reaching the steady state. The typical frequency transient behavior usually presents an overshoot, as shown in [22]. Hence, the maximum frequency deviation during the transient may be greater than the deviation in the steady state. This effect is not considered in [21].

1.2. Required Improvements in the EMS Development for Isolated Microgrids

As shown above, EMSs for isolated microgrids are commonly designed without analyzing their dynamic behavior. The primary and secondary control layers are responsible for stabilizing the microgrid after disturbance, but the EMS must consider their necessities to perform the operation properly. This issue has been previously solved by incorporating power reserves constraints in the optimization problems of the EMSs [14,19]. Nevertheless, few little dynamic considerations have been performed when designing EMSs. In addition to the power up/down regulation capacity, there are dynamic aspects that should be considered by the EMSs, which are not studied yet.

Utility grids are usually characterized by incorporating many rotating machines and, consequently, have large inertia. During power imbalances, and until the primary and secondary controls react, the required energy is obtained from the rotating machines leading to frequency variations. Due to the big inertia, these frequency variations are usually small. Accordingly, in grid connected microgrids, it can be assumed these deviations are not relevant [10]. In contrast, grid isolated microgrids present low inertia, and even lower when large amount of photovoltaic power is installed. Accordingly, these assumptions can no longer be accepted. Power reserves will determine whether the inner control loops can compensate the microgrid imbalances. However, due to the low inertia, the transient frequency deviations can reach unacceptable levels collapsing the system. Even though the study performed in [19] considers the up/down regulation and analyzes the dynamic response, a required inertia to ensure the frequency does not exceed the acceptable limit is not studied. Hence, in this case, the EMS developed in [19] disconnects to match the rotating machines, and the system stability could be compromised. Similarly, if frequency transients present overshoots, the stability of the system is not ensured by the proposed methods in [20,21].

According to the above issues, for designing a reliable EMS, it is still necessary to incorporate dynamic constraints into the problem formulation. Particularly, in addition to the power reserves, the minimum grid inertia to ensure an stable operation should be considered on the tertiary control layer of isolated microgrids.

1.3. Paper Contributions

This paper focuses on the above-mentioned issue. In particular, an EMS for ensuring that transient frequency deviations do not exceed a defined limit is developed. Accordingly, the main contribution of this paper are:

- The parameters that, being available by the EMS, may influence the frequency deviations are analyzed.
- The the maximum frequency deviation in front of the maximum power imbalance is formulated. This formulation uses the above-mentioned parameters.
- An EMS including a frequency constraint is formulated.
- The validation of the proposed EMS using dynamic simulation and laboratory platform is presented.

Particularly, this paper proposes a power dispatch optimization algorithm for PV–battery–diesel-based microgrids including demand and PV forecasting. To deal with uncertainty, the problem is based on stochastic optimization and computed on-line, in a similar way to the rolling horizon technique. The algorithm, which maximizes the PV generation, considers a frequency variation constraint obtained by analyzing multiple off-line dynamic simulations and performing a statistical study. The result shows that the minimum system frequency depends on the number of connected diesel generators, the battery power generation/consumption and the PV power generation. The algorithm was tested using simulation software (MATLAB-SIMULINK for simulation; and GAMS for solving the MILP optimization problem, using the SCIP solver) and validated in a laboratory platform. Particularly, three different days (based on real second-by-second data) were simulated. Then, one of the simulated days was tested in a laboratory-scaled microgrid platform.

2. System Description

The system under study is depicted in Figure 1. The microgrid consists of several diesel generators (N_d), where each unit i has a rated power P_{di}; a PV power plant, where the rated power is P_{pv-nom}; and a battery with rated power and capacity $P_{bat-nom}$ and C_{bat}, respectively. Finally, all these generation and storage units feed the total power demanded by the loads (P_c). The layout is based on a real stand alone system. It has the particularity that all generation and storage units (controllable units) are connected to the same bus. Thus, the load side can be treated as a single aggregated load. Each controllable unit has its local controller (LC) which is in charge of managing each resource separately:

- LC for diesel generation power plant: The local controller is in charge of controlling the frequency of the grid. A proportional–integral (PI) controller, where the input is the frequency error (filtered by a low pass filter), computes the mechanical torque setpoint of each diesel generator. This local controller also receives the required number of connected diesel generators and accordingly sends orders of connection/disconnection to each diesel unit. Each diesel generator has its internal controller in charge of reaching the torque setpoint and to perform its connection and disconnection according to the LC requirements. A similar control architecture is found in [23]. The main difference is that in the present paper the PI is a central controller that coordinates all the diesel units, while in [23] a single unit is considered.
- LC for the PV power plant: This LC implements a power–frequency droop curve to provide support to the grid. Reducing the active power will always be possible, but increasing it (under frequency events) will depend on the available active power. The controller can perform power curtailments. A maximum PV power setpoint is received externally and a PI controller computes the active power setpoint of each PV inverter. This controller is defined in [24], but the ramp rate limitation is not taken into account.
- LC for the battery: This controller receives externally an active power sepoint and applies a power–frequency droop curve to provide grid support. The output is the droop modified setpoint. The inner control loops will be in charge of reaching this value of active power. The dynamic model is simplified as in [25], but the local frequency droop has been included.

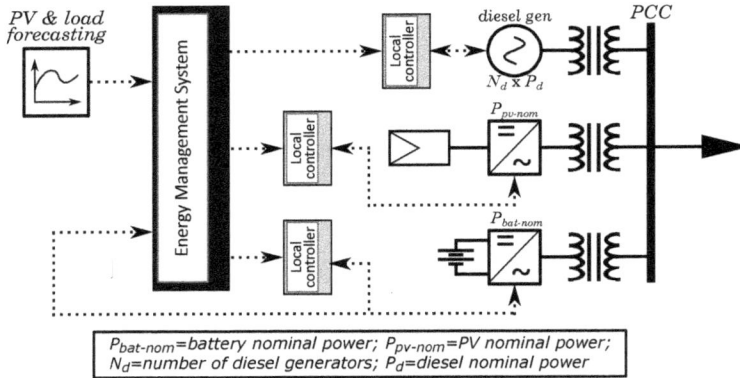

Figure 1. Simplified PV–battery–diesel-based microgrid scheme.

3. Methodology

3.1. EMS Design Requirements

The purpose of this section is to describe the steps followed for designing the EMS. The process is depicted in Figure 2. It shows that the EMS requirements are mainly determined by the characteristics of the system it will operate (system definition), the usage of the forecasting information (system data processing) and the identified operational requirements (system operation requirements).

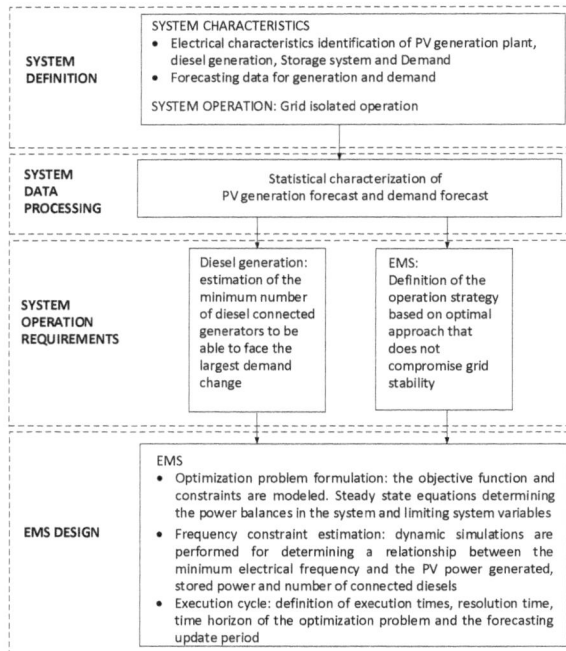

Figure 2. EMS design methodology.

First, the system characteristics are gathered—mainly the electrical characteristics and the available forecasting data—assuming grid isolated operation. Then, a statistical analysis of the forecasting for PV generation and demand is performed to identify the probability distribution of their errors. This allows generating random forecast scenarios (as detailed in Section 3.5). Next, the operation for the storage system is defined considering long-term variability of PV generation and demand. The minimum number of diesel generating units needed to face the largest demand change expected in the system is also determined. Finally, the EMS is designed, with two main purposes. On the one hand, the optimization problem is formulated based on the steady state equations determining the power balances in the system and limiting system variables. On the other hand, a frequency constraint, which is included in the optimization problem, is formulated (based on dynamic simulation results) relating the PV power generated, the battery power and the number of connected diesels with the minimum allowed frequency after a maximum power imbalance in the system.

The EMS performance is described in Section 3.2. The execution cycle of the EMS is detailed in Section 3.3. The procedure to determine the frequency constraint is explained in Section 3.4. For the stochastic optimization problem, it is required to generate a number of random scenarios, which is explained in Section 3.5. Finally, the whole optimization problem formulation is addressed in Section 3.7.

3.2. EMS Performance

The objective is to achieve the optimal utilization of the PV energy while achieving a generation–demand balance maintaining the grid frequency. In addition, it ensures that the minimum frequency (f^{mn}) reached after a severe generation–load imbalance is within the limits (see Section 3.4 and the frequency constraint explained later for more detail).

The output variables (the setpoints to the generation and storage units) of the EMS are: (i) number of diesel generators to be connected (D^*_{con}); (ii) the setpoint to the battery (P^*_{bat}); and (iii) the maximum PV power setpoint ($P^*_{PV_{max}}$). They are calculated for the remaining of the day at each optimization execution period. On the other hand, the inputs are: (i) the load forecast (L^c); (ii) the available PV power forecast (L^{PV}); and (iii) the initial state of charge (SOC). Forecasts include the mean and standard deviation.

Figure 3 shows the time periods used. (T_{for}) represents the time periods when forecasts are updated. (T_{EMS}) is the period between EMS executions. Finally, T_{intra} is the optimization problem time resolution. When the EMS is executed, the output variables (decision variables) are calculated for the rest of the day. While P^*_{bat} and P^*_{PV-max} are calculated with a time resolution of T_{intra}, the resolution of D^*_{con} is T_{EMS}.

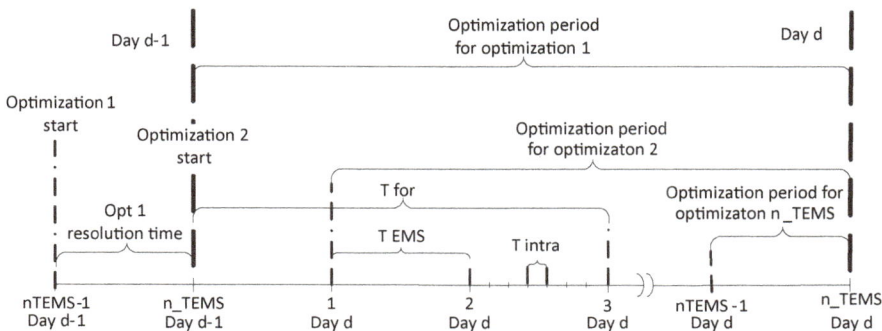

Figure 3. Temporal description of the daily execution cycle.

3.3. Execution Cycle

The optimization algorithm and its execution considers the daily sun period. Thus, the horizon of each execution is end of the day. This can be observed in Figure 3, where the execution cycle during the day d is depicted.

EMS period T execution: At period $T \in \{1, \dots, nT_{EMS}\}$, the $P_{T,p}^{bat^*}$ and $P_{T,p}^{PV_{max}^*} \ \forall \ p \in \{1, \dots, nT_{intra}\}$ are sent to their respective converters. These values are calculated in previous EMS executions (see Figure 3). Then, the SOC at the beginning of the EMS period $T + 1$ is estimated using the current SOC and the battery setpoints for the current execution period.

Using the estimated SOC at the EMS period $T + 1$ and the forecast for the rest of the day d, the optimization problem is solved, and $P_{t,p}^{bat^*}$ and $P_{t,p}^{PV_{max}^*} \ \forall \ p \in \{1, \dots, nT_{intra}\}$, $\forall t \in \{T + 1, \dots, nT_{EMS}\}$, and $D_t^{con^*} \ \forall \ t \in \{T + 1, \dots, nT_{EMS}\}$ are calculated.

The solution must be reached before the beginning of the EMS period $T + 1$. Otherwise, the setpoints calculated for the EMS period $T + 1$ by the EMS execution at the period $T - 1$ are sent to the respective converters.

3.4. Modeling Frequency Deviations

As explained above, one of the requirements of the isolated microgrid is the need to maintain the frequency in the required range. The frequency deviations depend on the grid inertia (i.e., the number of connected rotating machines) among other factors. One possible solution to ensure the frequency requirements is to connect the maximum number of rotating machines (diesel generators) providing large amount of inertia. However, these machines usually have a minimum active power generation. (The industry has reported that during low load condition diesel engines suffer from the "slobbering" effect. This effect is related to the low heat in the cylinder, allowing unburned fuel and oil to leak through the slip joints. This eventually leads to power losses, accelerated ageing and high maintenance costs). Thus, this strategy leads to a costly (fuel cost) and pollutant (CO_2 emissions) solution. Accordingly, the optimal solution is to connect the minimum number of rotating machines that ensures that, after a maximum power imbalance, the grid frequency will be kept in the required range.

Thus, the approach of this paper is to obtain an empirical linear equation determining the minimum frequency reached after a maximum power imbalance. This expression is then used in the optimization algorithm.

To obtain this expression, the worst case was first defined. The load and PV production of a real microgrid were monitored with 1 s resolution during six days and with 30 s resolution during one year. Using load data, a maximum load variation of 1.5 MW in 1 s was identified. This severe variation could have been produced due to the disconnection of a big load. For the case of PV data, it registered a maximum power variation of 1 MW in 1 s. According to the available recorded data, these changes will not occur simultaneously. Thus, the worst case considered is that the maximum power imbalance will occur after a sudden load variation of 1.5 MW, representing the situation when the maximum frequency deviation will occur.

Then, a simulation model of the microgrid was created. The model of the diesel generators are described in [23] while simplified PV and battery models are described in [25].

Using the simulation model, a bundle of scenarios varying D_{con} from N_d to N_{dmin} (being N_{dmin} the minimum number of diesel units connected to supply the maximum power imbalance), varying the P^{pv} from the rated PV power to 0 and varying the P^{bat} from P^{mxB} (maximum battery power) to P^{mnB} (minimum battery power) were simulated. In these simulations, the worst case (maximum load variation) was tested and the frequency response was analyzed, storing the minimum frequency reached for each simulation. From the analysis, a relation between the EMS output variables and the minimum frequency was performed (this analysis is explained below). To keep the optimization problem solvable using mixed integer linear programming (MILP), a linear regression is proposed for that purpose as Equation (1):

$$f^{mn} = \theta_{ind} + \theta_d \cdot D_{con} + \theta_{pv} \cdot P_{PV} + \theta_{bat} \cdot P_{bat} \tag{1}$$

where θ_x are the coefficients of linear regression.

The minimum frequency reached after the maximum power variation is represented in Figure 4 as a box plot against the ON^{dies}, P^{bat} and $P^{PV^*_{max}}$. For each of the decision variables, it is possible to observe the tendency of the minimum frequency reached. The lower is the P^{bat} and $P^{PV^*_{max}}$, the higher (in absolute values) is the maximum frequency deviation reached. On the other hand, the lower is the ON^{dies}, the lower is maximum frequency deviation reached. Figure 5 shows the summary of performing a linear regression; it can be observed that the coefficients for $P^{PV^*_{max}}$ and P^{bat} are negative and the coefficient for ON^{dies} is positive. The p-values for all the coefficients are lower than 10^{-8} and hence the obtained coefficients can be taken as significant.

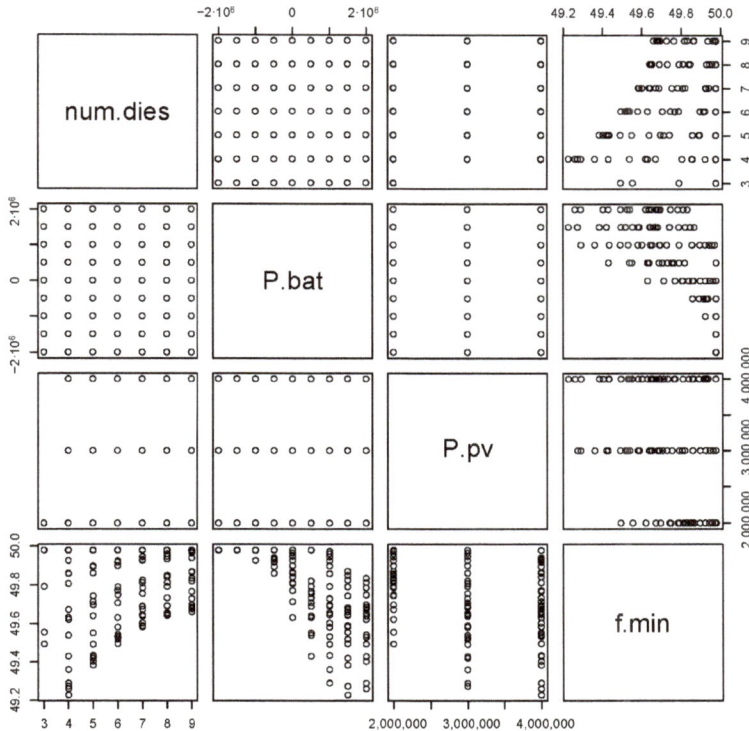

Figure 4. Box plot showing the relation between the minimum frequency reached and the decision variables of the EMS.

```
Coefficients:
               Estimate Std. Error  t value Pr(>|t|)
(Intercept)   4.991e+01  3.871e-02 1289.562  < 2e-16 ***
num.dies      2.723e-02  4.420e-03    6.160 5.24e-09 ***
P.bat        -1.129e-07  6.247e-09  -18.074  < 2e-16 ***
P.pv.curt    -8.798e-08  9.835e-09   -8.946 6.77e-16 ***
---
Signif. codes:  0 '***' 0.001 '**' 0.01 '*' 0.05 '.' 0.1 ' ' 1

Residual standard error: 0.1055 on 167 degrees of freedom
Multiple R-squared:  0.7218,    Adjusted R-squared:  0.7168
F-statistic: 144.5 on 3 and 167 DF,  p-value: < 2.2e-16
```

Figure 5. Linear regression results for the coefficients of the minimum frequency equation. Caption from R software.

3.5. Scenarios Generation

The forecasting system updates the forecasts for the rest of the day with a period T_{for}. The forecasts are based on a mean value and an error following a normal distribution with mean value ($\mu_{err} = 0$) and a standard deviation (σ_{err}). Using these values, the EMS generates a number of random scenarios N_s defined by the the the pair $L^c_{t,p,s}$ and $L^{PV}_{t,p,s}$, $\forall t \in \{To, \ldots, n_{TEMS}\}$, $\forall p \in \{1, \ldots, n_{Tintra}\}$; $\forall s \in \{1, \ldots, N_s\}$ being To the actual T^{EMS} period.

3.6. Stochastic Formulation Approach

The forecast errors are considered by using stochastic formulation. Particularly, N_s scenarios are generated (more details are given in Section 3.5). Then, the decision variables are constant for all scenarios, i.e., devices receive the same setpoints in all scenarios. In contrast, the rest of the variables are computed depending on each scenario. This way, the optimization problem ensures finding decision variables that fulfill the problem constraints for all scenarios generated. Then, the global objective function is the sum of the objective functions of all scenarios. Note that more probable scenario are generated more times and thus are counted more times in the global objective function, i.e., the most probable scenarios have higher weights in the objective function.

3.7. Formulation of the Optimization Algorithm

The optimization problem is stochastic. It means that, from the forecast (mean and deviation values), different scenarios are generated. The solution (the battery setpoints, the maximum PV power setpoints and the number of connected diesel generator setpoints) is unique independently of the scenario, but the constraints must be accomplished for all scenarios. The objective function is the sum of the objective functions of all scenarios. This way, we obtain an optimal solution considering forecast errors. In this section, the different optimization sets, decision variables and restrictions required to define the optimization problem are detailed.

3.7.1. Sets

The sets defining the EMS executions and the time resolution are shown in Equations (2) and (3), respectively.

$$T^{EMS} = \{1, \ldots, n_{TEMS}\} \tag{2}$$

$$T^{intra} = \{1, \ldots, n_{Tintra}\} \tag{3}$$

where n_{TEMS} is the number of the remaining executions of the optimization algorithm until the end of the day and n_{Tintra} is the number of periods of T_{intra} s between two executions of the optimization algorithm.

The index of the diesel generators are defined by the set in Equation (4), where N_d is the total number of diesel generators.

$$N^{diesel} = \{1, \ldots, N_d\} \tag{4}$$

Stochastic optimization is used to account for the forecast errors. Hence, each optimization execution considers N_s scenarios which are generated from the forecast inputs (mean and deviation). The set of the different scenarios is defined in Equation (5).

$$S = \{1, \dots, N_s\} \tag{5}$$

3.7.2. Decision Variables

The decision variables are those that the optimization algorithm finds to optimize the objective function.

The battery power setpoint is defined as Equation (6), where positive values of power mean that the battery is discharging. It is also distinguished if the battery is charging or discharging. The battery charging and discharging powers are defined as Equations (7) and (8), respectively. To prevent obtaining a solution where the battery could simultaneously charge and discharge, a binary variable is defined in Equation (9). The SOC is shown in Equation (10). In the EMS algorithm, it is assumed that the battery setpoint is the same as the real battery power generation/consumption.

$$P_{t,p}^{bat^*}, \ \forall\, t \in T^{EMS}, \ \forall\, p \in T^{intra} \tag{6}$$

$$P_{t,p}^{bat_{char}}, \ \forall\, t \in T^{EMS}, \ \forall\, p \in T^{intra} \tag{7}$$

$$P_{t,p}^{bat_{disch}}, \ \forall\, t \in T^{EMS}, \ \forall\, p \in T^{intra} \tag{8}$$

$$X_{t,p}^{char}, \ \forall\, t \in T^{EMS}, \ \forall\, p \in T^{intra}; X_{t,p}^{char} \in \{0,1\} \tag{9}$$

$$SOC_{t,p}^{bat}, \ \forall\, t \in T^{EMS}, \ \forall\, p \in T^{intra} \tag{10}$$

The diesel connection/disconnection setpoint and the power generation of each diesel generator are denoted as Equations (11) and (12), respectively. $ON_{t,p}^{dies}$ is 1 if the diesel generator d at the EMS period t and the intra period d is connected (and 0 otherwise).

$$ON_{t,p}^{dies}, \ \forall\, t \in T^{EMS}, \ \forall\, d \in N^{Diesel}, \ ON_{t,p}^{dies} \in \{0,1\} \tag{11}$$

$$P_{t,p,s,d}^{dies}, \ \forall\, d \in N^{Diesel}, \ \forall\, s \in S, \ \forall\, t \in T^{EMS}, \ \forall\, p \in T^{intra} \tag{12}$$

The PV power generation of each scenario is written as Equation (13), while the maximum PV power setpoint is expressed as Equation (14).

$$P_{t,p,s}^{pv}, \ \forall\, s \in S, \ \forall\, t \in T^{EMS}, \ \forall\, p \in T^{intra} \tag{13}$$

$$P_{t,p}^{PV_{max}^*}, \ \forall\, t \in T^{EMS}, \ \forall\, p \in T^{intra} \tag{14}$$

3.7.3. Parameters

The load and PV scenarios are generated according to the forecast mean values and deviations. These scenarios are expressed as Equations (15) and (16), respectively. They represent the active power of load and the available PV power.

$$L_{t,p,s}^{c}, \ \forall\, s \in S, \ \forall\, t \in T^{EMS}, \ \forall\, p \in T^{intra} \tag{15}$$

$$L_{t,p,s}^{PV}, \ \forall\, s \in S, \ \forall\, t \in T^{EMS}, \ \forall\, p \in T^{intra} \tag{16}$$

The battery capacity, the initial SOC, the battery efficiency and the maximum and minimum battery active power are written as Cap^{bat}, SOC^i, η^{bat}, P^{mxB} and P^{mnB}, respectively. The maximum and minimum active power of each diesel unit are expressed as P^{mxD} and P^{mxD}, respectively. The minimum frequency is expressed as f^{mn}. The diesel generators performs a frequency control through a PI controller. To provide a power reserve for frequency regulation, a power margin of diesel generators is reserved. This power margin is denoted as $marge_{dies}$.

3.7.4. Objective Function

The objective function it to maximize the PV power generation. To do so, the battery is charged or discharged according to the forecast and the problem requirements. During the charging and discharging process, there are some power losses. Thus, the real useful PV power must take into account them. Accordingly, the objective function is written as Equation (17).

$$[MAX]\, Z = \sum_{t,p,s} P^{PV}_{t,p,s} - n_S\, (1 - \eta_{bat})\, abs(P^{bat^*}_{t,p}) \tag{17}$$

To linearize this function, it can be re-written as Equation (18).

$$[MAX]\, Z = \sum_{t,p,s} P^{PV}_{t,p,s} - n_S\, (1 - \eta_{bat})\, (P^{bat_{char}}_{t,p} + P^{bat_{disch}}_{t,p}) \tag{18}$$

3.7.5. Constraints

The objective function has been linearized but. to prevent obtaining simultaneous charge and discharge of the battery, the following constrains are included (Equations (19)–(23)):

$$P^{bat^*}_{t,p} = P^{bat_{char}}_{t,p} - P^{bat_{disch}}_{t,p}\ \forall\, t \in T^{EMS},\ \forall\, p \in T^{intra} \tag{19}$$

$$P^{bat_{char}}_{t,p} \le P^{mxB} X^{char}_{t,p}\ \forall t \in T^{EMS},\ \forall\, p \in T^{intra} \tag{20}$$

$$P^{bat_{disch}}_{t,p} \le P^{mxB}(X^{char}_{t,p} - 1)\ \forall\, t \in T^{EMS},\ \forall\, p \in T^{intra} \tag{21}$$

$$P^{bat_{char}}_{t,p} \ge 0\ \forall\, t \in T^{EMS},\ \forall\, p \in T^{intra} \tag{22}$$

$$P^{bat_{disch}}_{t,p} \ge 0\ \forall\, t \in T^{EMS},\ \forall\, p \in T^{intra} \tag{23}$$

Then, the power balance at each period must be accomplished. This is forced by the restriction in Equation (24).

$$P^{pv}_{t,p,s} + \sum_{d \in N^{Diesel}} P^{dies}_{t,p,s,d} + P^{bat^*}_{t,p} - L^{c}_{t,p,s} = 0 \forall\, t \in T^{EMS},\ \forall\, p \in T^{intra},\ \forall\, s \in S \tag{24}$$

Then, as commented above, a margin of diesel generation is reserved for frequency regulation. Thus, the maximum diesel generation is limited (Equation (25)):

$$\sum_{d \in N^{Diesel}} P^{dies}_{t,p,s,d} \le \sum_{d \in N^{Diesel}} ON^{dies}_{t,d} P^{mxD} - marge_{dies} \forall\, t \in T^{EMS},\ \forall\, p \in T^{intra},\ \forall\, s \in S \tag{25}$$

The relationship between the SOC at instant t and the SOC at instant $t-1$ is shown in Equation (26). The SOC is between 0 and 1 p.u. This constraint is formulated as Equation (27). On the other hand, the battery power limits constraint is Equation (28).

$$
\begin{aligned}
&\text{-If } T^{EMS} = 1 \text{ and } T^{intra} = 1 \\
&SOC_{t,p}^{bat} = SOC^{initial} - P_{t,p}^{bat}\frac{\Delta t}{Cap^{bat}} \; \forall\, t \in T^{EMS}, \; \forall\, p \in T^{intra} \\
&\text{-If } T^{EMS} \geq 1 \text{ and } T^{intra} = 1 \\
&SOC_{t,p}^{bat} = SOC_{t-1,|p|}^{bat} - P_{t,p}^{bat}\frac{\Delta t}{Cap^{bat}} \; \forall\, t \in T^{EMS}, \; \forall\, p \in T^{intra} \\
&\text{-If } T^{intra} \neq 1 \\
&SOC_{t,p}^{bat} = SOC_{t,p-1}^{bat} - P_{t,p}^{bat}\frac{\Delta t}{Cap^{bat}} \; \forall\, t \in T^{EMS}, \; \forall\, p \in T^{intra}
\end{aligned}
\tag{26}
$$

$$
0 \leq SOC_{t,p}^{bat} \leq 1 \; \forall\, t \in T^{EMS}, \; \forall\, p \in T^{intra}
\tag{27}
$$

$$
P^{mnB} \leq P_{t,p}^{bat} \leq P^{mxB} \; \forall\, t \in T^{EMS}, \; \forall\, p \in T^{intra}
\tag{28}
$$

Then, the PV power cannot be greater than the available PV power of the corresponding scenario. Thus, Equation (29) must be included into the optimization algorithm. The PV power must also be lower than the maximum PV power setpoint of Equation (30).

$$
P_{t,p,s}^{PV} \leq L_{t,p,s}^{PV} \; \forall\, t \in T^{EMS}, \; \forall\, p \in T^{intra}, \; \forall\, s \in S
\tag{29}
$$

$$
P_{t,p,s}^{PV} \leq P_{t,p}^{PV_{max}} \; \forall\, t \in T^{EMS}, \; \forall\, p \in T^{intra}, \; \forall\, s \in S
\tag{30}
$$

Each diesel unit has a maximum and a minimum power at each scenario, which is formulated as Equation (31).

$$
ON_{d,t,s}^{dies} P_d^{mnD} \leq P_{t,p,s}^{dies} \leq P^{mxD} ON_{d,t,s}^{dies} \forall\, t \in T^{EMS}, \forall\, p \in T^{intra}, \; \forall\, s \in S
\tag{31}
$$

Finally, the minimum frequency constraint is included in the optimization model. In the previous section, it has been shown how to express the minimum frequency reached in the microgrid after a maximum power imbalance. This constraint is written as Equation (32).

$$
f^{mn} \leq \theta_{ind} + \theta_d \sum_d ON_{t,d}^{dies} + \theta_{bat} P_{t,p}^{bat^*} + \theta_{pv} P_{t,p}^{PV_{max}^*} \forall\, t \in T^{EMS}, \; \forall\, p \in T^{intra}
\tag{32}
$$

4. Case Study

Based on a real case, the microgrid included: 9×1.2 MVA diesel units and 2×560 kWh batteries, which are interconnected through 4×550 kVA inverters (2 inverters per battery). The total battery power was then 2.2 MVA. The rated power of the PV plant was 10 MW, similar to the one presented in [24]. The minimum accepted frequency was $f^{mn} = 49.0$ Hz. Table 1 shows the problem parameters.

Table 1. Parameters for the EMS optimization problem.

Parameter	Value	Parameter	Value	Parameter	Value
n_{TEMS}	288	Cap^{bat}	1120 kWh	P^{mnB}	-2200 kW
n_{Tintra}	10	SOC^i	0.9	P^{mnD}	$0.3 \cdot 1100$ kW
N_d	9	η^{bat}	0.9	P^{mxD}	1100 kW
N_s	5	P^{mxB}	2200 kW	$marge_{dies}$	2000 kW

Three scenarios were simulated. The load consumption was the same for all scenarios and shown in the result plots. The difference between the three scenarios was reflected in the available PV power profile. In the first case, after 12:30, the available PV profile presented large variations. The second scenario had lower PV variability, but it was not a fully sunny day. Finally, the last case consisted of a sunny day with not appreciably fast PV power variations. The simulation results are shown in Figure 6 for the first case, in Figure 7 for the second case and in Figure 8 for the last case. Note that the simulation considered the execution cycle explained in Section 3.3 and the EMS outputs were introduced to the dynamic model.

For each scenario, the top plot depicts the active power of microgrid's devices as well as the power demand and the available PV power. In the middle plot, the SOC and the connections of diesel units can be observed. Then, the bottom plot shows the frequency response of the microgrid in blue, being the green lines the frequency droop dead-band (out of this range, the PV plant and the batteries provide frequency support). It can be observed that, for the three scenarios, the battery was discharged at the beginning of the day to be able to charge during the hours of high PV power. In addition, as could be expected, the active power of diesel generators and the connected units followed a trend complementary to the PV power generation. Thus, during the peak PV production hours, the amount of connected diesel generators was lower, as well as their production. It is also shown that the frequency deviations were kept inside the acceptable range. Comparing the total PV energy generated to the available PV energy for the three scenarios, the relative amount of used PV energy was 94.57%, 84.46% and 94.98%, respectively. The second scenario had the lowest PV profitability, but note that, in this case, the maximum available PV power was higher than the load in some periods.

During the times 13:00–15:00, the frequency exceeded the droop dead-band several times. Thus, the PV and battery provided frequency support. This happens because, during this period, the number of connected diesel generators was small (low inertia). Hence, either the large PV fluctuations or the connection of new generators injecting active power produced a frequency transient. While the frequency may have exceeded the frequency droop dead-band (green lines), it did not exceed the minimum value of 49 Hz.

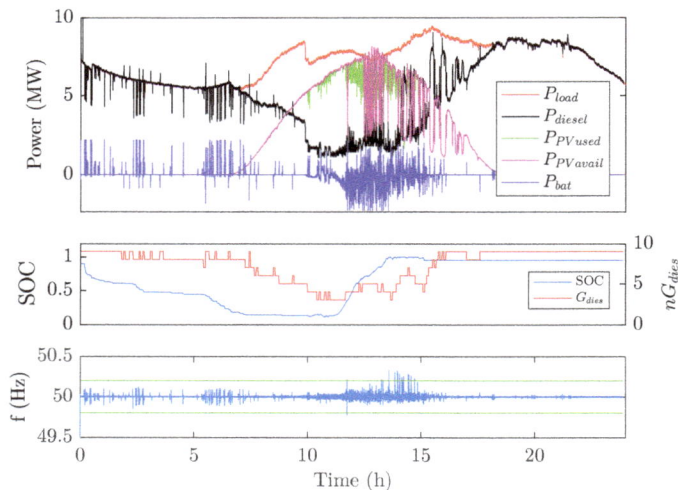

Figure 6. Simulation results for the first scenario (high PV power variability after the midday).

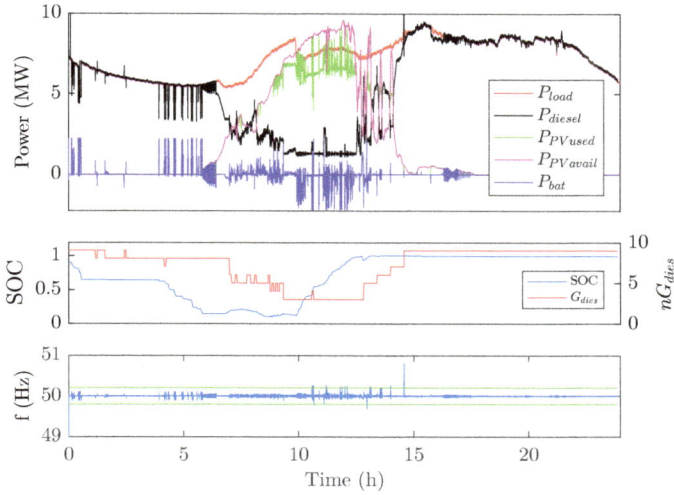

Figure 7. Simulation results for the second scenario (medium PV variability).

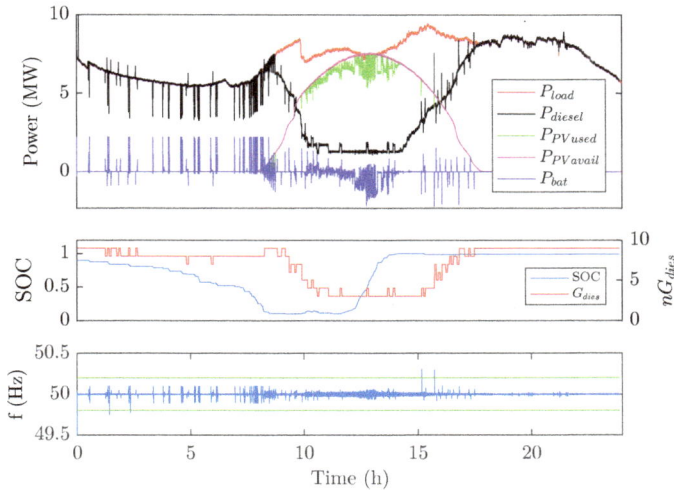

Figure 8. Simulation results for the third scenario (low PV variability).

5. Experimental Validation

5.1. Platform Description

An emulated microgrid was used for the experiments. As described in [26], an emulator consists of a platform able to convert software processed variables into real magnitudes. Accordingly, real equipment can be tested by its interconnection to the emulator platform. Hence, the system presented above can be tested properly through the emulation concept.

The layout of the laboratory microgrid (emulated microgrid) and its physical devices are depicted in Figure 10a,b, respectively. The emulated devices (diesel units, PV generators, storage, and loads) mimicked the behavior of the real device they are representing and form the emulated subsystem of the experimental setup. They were configured using a dedicated PC and a communication network. On the other hand, the real devices of the experimental setup were the PV and battery inverters, the power transformers, the EMS (which was implemented on a dedicated PC) as well as the communication network and the SCADA system. Because it is desired to emulate the isolated operation, the switch interconnecting the real system with the external grid was opened.

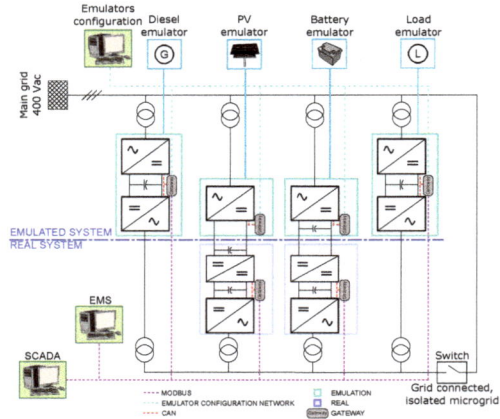

(**a**) Microgrid emulator scheme

Figure 9. *Cont.*

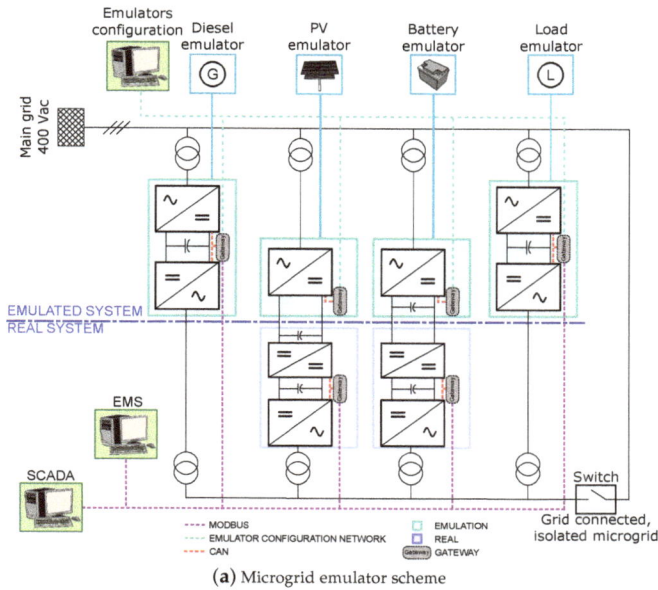

(**a**) Microgrid emulator scheme

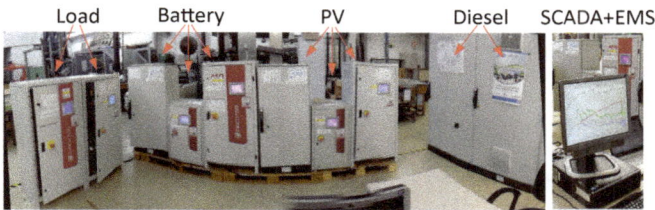

(**b**) Microgrid photo

Figure 10. Microgrid description.

5.2. Emulation Results

The simulated results were validated using the first test case (the one presenting the highest PV power variability) and the emulation platform under a real time emulation test. The input data were scaled-down according the emulator's power ratings. The outputs of the EMS were sent, periodically (T_{EMS} = 5 min), to the devices (emulated). In Figure 11, the experimental results can be observed, showing how the response was very similar to the simulation results. In particular, the same tendency in the diesel units connections and disconnections as well as in the battery utilization can be observed. An important observation is that, generally, the generation was greater than the load because the emulated inverters had power losses.

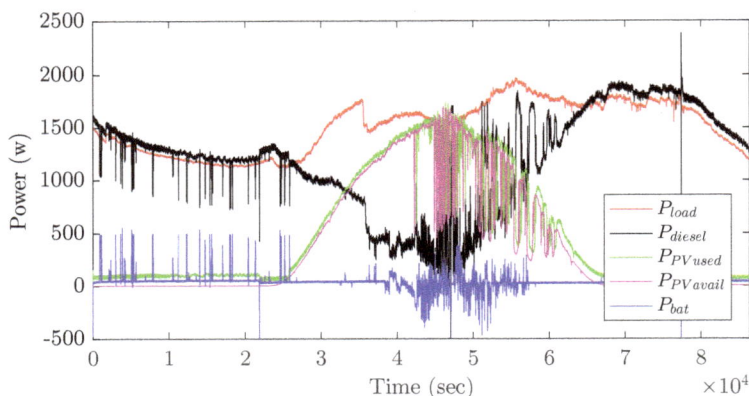

Figure 11. Laboratory emulation results for the first scenario (high PV power variability after the midday).

6. Conclusions

A new methodology for the optimal operation of isolated microgrids has been proposed. This methodology is based on stochastic optimization to consider the forecast errors. In addition, a minimum frequency constraint has been formulated and included the optimization algorithm to ensure the secure operation of the microgrid. To maintain the optimization problem as a mixed integer linear problem, this constraint has been defined using a linear regression.

Three different scenarios, based on real data, were tested using a dynamic model of the microgrid. The results show a good behavior with a stable grid frequency and high rate of PV energy used.

After proving the proper response of the EMS using a simulation model, it was implemented to manage a laboratory scale microgrid, where real time limitations, communication delays and measurement errors occur. It has been shown that the system can also operate properly with real platforms having similar behavior to the simulated system.

Author Contributions: Conceptualization, J.-A.V.-C., E.B.-M., M.A.-P., and G.V.-C.; methodology, J.-A.V.-C., E.B.-M., M.A.-P., and G.V.-C.; software, J.-A.V.-C., C.C.-A. and E.P.-A.; validation, J.-A.V.-C., E.B.-M. and M.A.-P.; formal analysis, J.-A.V.-C.; investigation, J.-A.V.-C. and G.V.-C.; writing–original draft preparation, J.-A.V.-C., E.B.-M. and M.A.-P.; writing–review and editing, all authors; supervision, O.G.-B. and S.G.-A.

Funding: This research received no external funding

Acknowledgments: The research leading to these results received support of the Secretaria d'Universitats i Recerca del Departament d'Economia i Coneixement de la Generalitat de Catalunya. M. Aragüés-Peñalba and E. Prieto-Araujo are lecturers of the Serra Húnter Programme. The authors would like to thank Luis Serrano and Carlos Pacheco from GreenPowerMonitor for their support in this work.

Conflicts of Interest: The authors declare no conflict of interest.

References

1. Aegerl, C.S.; Tao, L. The Microgrids Concept. In *Microgrids Architectures and Control*; Hatziargyriou, N., Ed.; John Wiley and Sons: Hoboken, NJ, USA, 2013.
2. DOE. *Summary Report: 2012 DOE Microgrid Workshop*; DOE: Washington, DC, USA, 2012.
3. Bullich-Massagué, E.; Díaz-González, F.; Aragüés-Peñalba, M.; Girbau-Llistuella, F.; Olivella-Rosell, P.; Sumper, A. Microgrid clustering architectures. *Appl. Energy* **2018**, *212*, 340–361. [CrossRef]
4. Neely, J.; Johnson, J.; Delhotal, J.; Gonzalez, S.; Lave, M. Evaluation of PV frequency-watt function for fast frequency reserves. In Proceedings of the 2016 IEEE Applied Power Electronics Conference and Exposition (APEC), Long Beach, CA, USA, 20–24 March 2016; pp. 1926–1933. [CrossRef]
5. Kundur, P.; Balu, N.; Lauby, M. *Power System Stability and Control*; EPRI Power System Engineering Series; McGraw-Hill: New York, NY, USA, 1994.

6. Martin-Martínez, F.; Sánchez-Miralles, A.; Rivier, M. A literature review of Microgrids: A functional layer based classification. *Renew. Sustain. Energy Rev.* **2016**, *62*, 1133–1153. [CrossRef]
7. Han, Y.; Shen, P.; Zhao, X.; Guerrero, J.M. Control Strategies for Islanded Microgrid Using Enhanced Hierarchical Control Structure With Multiple Current-Loop Damping Schemes. *IEEE Trans. Smart Grid* **2017**, *8*, 1139–1153. [CrossRef]
8. Vandoorn, T.L.; Vasquez, J.C.; Kooning, J.D.; Guerrero, J.M.; Vandevelde, L. Microgrids: Hierarchical Control and an Overview of the Control and Reserve Management Strategies. *IEEE Ind. Electron. Mag.* **2013**, *7*, 42–55. [CrossRef]
9. Bidram, A.; Davoudi, A. Hierarchical Structure of Microgrids Control System. *IEEE Trans. Smart Grid* **2012**, *3*, 1963–1976. [CrossRef]
10. Parisio, A.; Rikos, E.; Glielmo, L. A Model Predictive Control Approach to Microgrid Operation Optimization. *IEEE Trans. Control Syst. Technol.* **2014**, *22*, 1813–1827. [CrossRef]
11. Marzband, M.; Sumper, A.; Domínguez-García, J.L.; Gumara-Ferret, R. Experimental validation of a real time energy management system for microgrids in islanded mode using a local day ahead electricity market and MINLP. *Energy Convers. Manag.* **2013**, *76*, 314–322. [CrossRef]
12. Marzband, M.; Yousefnejad, E.; Sumper, A.; Domínguez-García, J.L. Real time experimental implementation of optimum energy management system in standalone Microgrid by using multi layer ant colony optimization. *Int. J. Electr. Power Energy Syst.* **2016**, *75*, 265–274. [CrossRef]
13. Marzband, M.; Ghadimi, M.; Sumper, A.; Domínguez-García, J.L. Experimental validation of a real-time energy management system using multi period gravitational search algorithm for microgrids in islanded mode. *Appl. Energy* **2014**, *128*, 164–174. [CrossRef]
14. Chen, C.; Duan, S.; Cai, T.; Liu, B.; Hu, G. Smart energy management system for optimal microgrid economic operation. *IET Renew. Power Gener.* **2011**, *5*, 258–267. [CrossRef]
15. Sobu, A.; Wu, G. Optimal operation planning method for isolated micro grid considering uncertainties of renewable power generations and load demand. In Proceedings of the IEEE PES Innovative Smart Grid Technologies, Tianjin, China, 21–24 May 2012; pp. 1–6. [CrossRef]
16. Lazaroiu, G.C.; Dumbrava, V.; Balaban, G.; Longo, M.; Zaninelli, D. Stochastic optimization of microgrids with renewable and storage energy systems. In Proceedings of the 2016 IEEE 16th International Conference on Environment and Electrical Engineering (EEEIC), Florence, Italy, 7–10 June 2016; pp. 1–5. [CrossRef]
17. Cau, G.; Cocco, D.; Petrollese, M.; Kær, S.K.; Milan, C. Energy management strategy based on short-term generation scheduling for a renewable microgrid using a hydrogen storage system. *Energy Convers. Manag.* **2014**, *87*, 820–831. [CrossRef]
18. Palma-Behnke, R.; Benavides, C.; Lanas, F.; Severino, B.; Reyes, L.; Llanos, J.; Sáez, D. A Microgrid Energy Management System Based on the Rolling Horizon Strategy. *IEEE Trans. Smart Grid* **2013**, *4*, 996–1006. [CrossRef]
19. Zhao, B.; Xue, M.; Zhang, X.; Wang, C.; Zhao, J. An {MAS} based energy management system for a stand-alone microgrid at high altitude. *Appl. Energy* **2015**, *143*, 251–261. [CrossRef]
20. Sanseverino, E.R.; Nguyen, N.Q.; Silvestre, M.L.D.; Zizzo, G.; de Bosio, F.; Tran, Q.T.T. Frequency constrained optimal power flow based on glow-worm swarm optimization in islanded microgrids. In Proceedings of the 2015 AEIT International Annual Conference (AEIT), Naples, Italy, 14–16 October 2015; pp. 1–6. [CrossRef]
21. Zhang, G.; McCalley, J. Optimal power flow with primary and secondary frequency constraint. In Proceedings of the 2014 North American Power Symposium (NAPS), Pullman, WA, USA, 7–9 September 2014; pp. 1–6. [CrossRef]
22. Díaz-González, F.; Hau, M.; Sumper, A.; Gomis-Bellmunt, O. Participation of wind power plants in system frequency control: Review of grid code requirements and control methods. *Renew. Sustain. Energy Rev.* **2014**, *34*, 551–564. [CrossRef]
23. Theubou, T.; Wamkeue, R.; Kamwa, I. Dynamic model of diesel generator set for hybrid wind-diesel small grids applications. In Proceedings of the 2012 25th IEEE Canadian Conference on Electrical and Computer Engineering (CCECE), Montreal, QC, Canada, 29 April–2 May 2012; pp. 1–4. [CrossRef]
24. Bullich-Massagué, E.; Ferrer-San-José, R.; Aragüés-Peñalba, M.; Serrano-Salamanca, L.; Pacheco-Navas, C.; Gomis-Bellmunt, O. Power plant control in large-scale photovoltaic plants: Design, implementation and validation in a 9.4 MW photovoltaic plant. *IET Renew. Power Gener.* **2016**, *10*, 50–62. [CrossRef]

25. Bullich-Massagué, E.; Aragüés-Peñalba, M.; Sumper, A.; Boix-Aragones, O. Active power control in a hybrid PV-storage power plant for frequency support. *Sol. Energy* **2017**, *144*, 49–62. [CrossRef]

26. Prieto-Araujo, E.; Olivella-Rosell, P.; Cheah-Mañe, M.; Villafafila-Robles, R.; Gomis-Bellmunt, O. Renewable energy emulation concepts for microgrids. *Renew. Sustain. Energy Rev.* **2015**, *50*, 325–345. [CrossRef]

![applied sciences logo] *applied sciences*

MDPI

Article

Hierarchical Optimization Method for Energy Scheduling of Multiple Microgrids

Tao Rui [1,2], Guoli Li [2], Qunjing Wang [2], Cungang Hu [2,*], Weixiang Shen [3] and Bin Xu [4]

[1] School of Computer Science and Technology, Anhui University, Hefei 230601, China; ruitao5555@163.com
[2] Engineering Research Center of Power Quality, Ministry of Education, Anhui University, Hefei 230601, China; liguoli@ahu.edu.cn (G.L.); wqunjing@sina.com (Q.W.)
[3] Faculty of Science, Engineering and Technology, Swinburne University of Technology, Melbourne 3122, Australia; wshen@swin.edu.au
[4] State Grid Anhui Electric Power Co. Ltd. Research Institute, Hefei 230601, China; xubin1980@sina.com
* Correspondence: hcg@ahu.edu.cn; Tel.: +86-158-5511-5115

Received: 29 December 2018; Accepted: 11 February 2019; Published: 13 February 2019

Abstract: This paper proposes a hierarchical optimization method for the energy scheduling of multiple microgrids (MMGs) in the distribution network of power grids. An energy market operator (EMO) is constructed to regulate energy storage systems (ESSs) and load demands in MMGs. The optimization process is divided into two stages. In the first stage, each MG optimizes the scheduling of its own ESS within a rolling horizon control framework based on a long-term forecast of the local photovoltaic (PV) output, the local load demand and the price sent by the EMO. In the second stage, the EMO establishes an internal price incentive mechanism to maximize its own profits based on the load demand of each MG. The optimization problems in these two stages are solved using mixed integer programming (MIP) and Stackelberg game theory, respectively. Simulation results verified the effectiveness of the proposed method in terms of the promotion of energy trading and improvement of economic benefits of MMGs.

Keywords: multiple microgrid; rolling optimization; Stackelberg game; price mechanism

1. Introduction

With the increasing penetration of renewable energy resources (RESs) in the distribution networks of regional power grids, intermittent RESs, which are connected to distribution networks in a distributed way with a small capacity and high density, have great influence on the stability of regional power grids. The energy management and control of distributed generation, load demand and energy storage systems (ESSs) using microgrid (MG) technology can effectively improve the stability of regional power grids [1–3]. MGs with different characteristics coexist in distribution networks of regional power grids and form multiple microgrids (MMGs). MGs with different characteristics can achieve energy interaction through energy management systems (EMSs), which not only can enhance the reliability of the regional power supply [4], but also promote the utilization of renewable energy, and improve the economic benefits of MGs [5].

In order to manage the energy scheduling between different MGs, centralized coordination control is commonly adopted in MMG architecture [6–8]. In [9], the energy market operator (EMO) acts as a centralized control system to coordinate the energy iteration between a cluster of selling MGs and a cluster of buying MGs. A two-stage robust optimization for energy transactions in MMGs is proposed in [10], which could minimize system cost under the worst realization of uncertain PV output. In [11], a multiple agent system (MAS)-based hierarchical energy management strategy for MMGs is proposed with easy implementation and low computation cost. A practical model is proposed for distribution companies to minimize the total operation cost of the system including distribution networks and

MGs through coordinated operation [12]. However, the above optimization methods are all aimed at a situation in which there is no direct conflict of interest between MGs and upper manager. And these methods are not very suitable for competitive hierarchical power market structures [13].

To stimulate the energy transaction potential of MGs, a multi-market participation framework is proposed for distribution network operators (DNOs) [14]. Since market participants belong to different stakeholders, DNOs and MGs can trade energy not only through cooperation, but also through competition. The relationship among participants is suitable to be solved by game theory. In [15], an incentive mechanism based on cooperative game theory is proposed to reduce the peak ramp of distribution networks and improve the benefits of MGs. In [16], a two-level day-ahead scheduling structure based on Stackelberg game is proposed to stimulate MGs participating in power sale bidding, which could reduce the total cost of the DNO. A bi-level programming based on Stackelberg is adopted by the DNO to improve the benefits of MGs while the cost of the DNO reaches the minimum [17]. However, these methods are based on day-head energy market, and there are still some shortcomings in dealing with the uncertainties of renewable energy generation and load demand.

In this paper, a two-level optimization method is proposed for hour-ahead MMG energy scheduling in distribution network energy markets, where the EMO is the upper manager of the whole MMG system, and the MGO is lower manager of the local MG. This method can be implemented in two stages. According to the short-term forecast information, the local MGO optimizes energy scheduling of storage units by adopting rolling optimization in the first stage. The energy transactions between the EMO and MGs are optimized in the second stage, which is modeled as a Stackelberg game. The EMO is the leader of the game who determines the prices of the next hour to maximize its own utility, while the MGs are followers who respond to prices by adjusting local load demands. Through two stages of optimization, the economic utilities of the EMO and MGOs are both improved. The other significant feature of this proposed method is that energy storage and load demands are dispatched hourly, which makes it more reasonable to manage energy trading in MMGs under the uncertainties of PV output and load demands.

2. Framework

2.1. Structure of an MMG

The structure of an MMG in this paper is shown in Figure 1. The parties involved in energy trading in the MMG are MGs, the EMO and a power grid. Each MG mainly consists of a PV system, an ESS, loads, a smart meter and a local EMS. For a rational MG, the first choice of PV generation is to supply its own loads and charge its own ESS. The MG can then act as a seller when there is an energy surplus or as a buyer when there is an energy deficit. The net energy profile of the MG is optimized by the local EMS according to the benefits generated by energy consumption and the internal prices from the EMO. The EMO is responsible for stimulating energy trading in MMGs by establishing a reasonable internal buying price (p_{cb}^h) and selling price (p_{cs}^h) each hour. At the occurrence of an internal energy mismatch, the EMO trades with power grids to balance supply and demand. In order to guarantee that the profits of the MG obtained from energy sharing among MGs are better than those of the MG obtained from energy trading with power grids directly, the internal prices produced by the EMO should be between the selling price (p_{gs}^h) and the buying price (p_{gb}^h) of power girds, allowing the EMO to maximize its own profits under the constraint as follows:

$$p_{gs}^h \leq p_{cs}^h < p_{cb}^h \leq p_{gb}^h \tag{1}$$

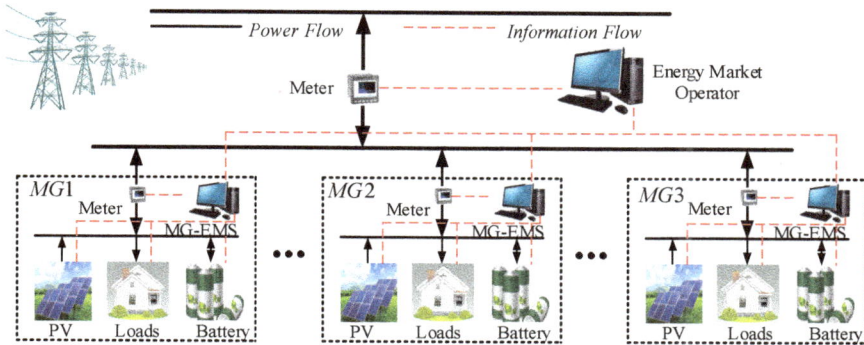

Figure 1. Structure of an MMG (multiple microgrid).

2.2. Operation Strategy

Figure 2 shows the details of the operation strategy of an MMG, which can be described by two stages of optimization.

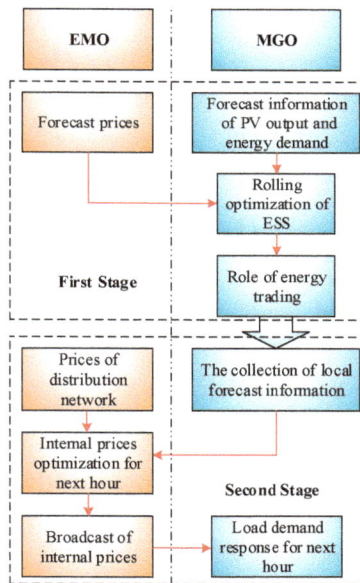

Figure 2. Operation strategy of an MMG (EMO: Energy Market Operator, MGO: Microgrid Operator, PV: photovoltaic).

2.2.1. First Stage

Considering the time-coupling characteristics of a local ESS, a local EMS adopts a rolling optimization method to determine the scheduling plan of the ESS for next period, which can minimize the cost for the MG to purchase energy from the EMO and determine the role of the MG to participate in energy trading in the next period. The inputs of the rolling optimization are the long-term forecast information of local PV output, local energy demand and the price obtained from the EMO.

2.2.2. Second Stage

According to the short term forecast information collected from MGs and the prices of power grids, the EMO establishes the internal price optimization model for next period based on Stackelberg game theory. The optimization target of the EMO is to maximize its own profits while considering the demand responses of MGs. Furthermore, the EMO broadcasts the results of the internal prices to the MGs, which respond to the prices by adjusting load demands.

3. System Model

3.1. Utility Model of MGs

Each MG equipped with a PV system receives a government subsidy for clean energy plus the revenue from selling surplus energy to MMGs. The MG prefers the PV generation to satisfy its own load demands and charge its own ESS. If the PV generation is insufficient, the MG will buy energy form the MMG. The energy consumption in the MG can create revenue, especially for industrial and commercial users [18,19]. Therefore, the utility model of the MG mainly considers the government subsidy for PV generation, the benefits of energy consumption, the benefits from selling surplus energy to the EMO and the costs of purchasing energy from the EMO. As the MG may be either a buyer or a seller in different periods, the utility function in time slot h can be expressed as follows:

$$U_i^h = \theta p v_i^h + k_i^h \ln(+1 l_i^h) - p_{cb}^h n l_{s,i}^h - p_{cs}^h n l_{b,i}^h - c_i e_i^{h2} \tag{2}$$

where θ is the subsidy for each kWh generated by the PV system, pv_i^h is the PV generation in the time slot h. $k_i^h \ln(1 + l_i^h)$ is the benefit that the MG i consumes energy l_i^h. p_{cb}^h and p_{cs}^h are the buying price and selling price from the MG to the MMG, respectively. $nl_{s,i}^h$ and $nl_{b,i}^h$ respectively represent the selling energy and buying energy of the MG i in time slot h. c_i is the degradation cost coefficient of the ESS in the MG i, and e_i^h is the charging energy to the ESS in time slot h. The other maintenance and scheduling costs are neglected in this paper. Since each MG in the MMG can only act as a buyer or seller at a special time slot, and the net energy should satisfy the following constraints:

$$nl_{b,i}^h - nl_{s,i}^h = l_i^h + e_i^h - pv_i^h, \tag{3}$$

$$0 \leq nl_{b,i}^h \leq D_{b,i}^h nl_i^{\max}, \tag{4}$$

$$0 \leq nl_{s,i}^h \leq D_{s,i}^h nl_i^{\max}, \tag{5}$$

$$D_{b,i}^h + D_{s,i}^h \leq 1. \tag{6}$$

Equation (3) is the constraint of energy balance in the MG i. $nl_i^{\max}$ is the maximum energy transaction of the MG with the power grid in the connection line, $D_{b,i}^h$ and $D_{s,i}^h$ are the binary variables indicating the state of buying and selling energy of the MG i.

3.2. Profit Model of EMOs

The MG may be either a seller or a buyer depending on its requirement of net energy. The MMG energy importing from the MGs or exporting to the MGs in time slot h can be expressed by

$$E_{im}^h = \sum_{D_{s,i}^h = 1} nl_{s,i}^h, \tag{7}$$

$$E_{ex}^h = \sum_{D_{b,i}^h = 1} nl_{b,i}^h. \tag{8}$$

Usually, the mismatch between imported energy and exported energy always exists, and the EMO should trade with the power grid to maintain an internal energy balance. Therefore, the profit function of the EMO can be written as

$$Pro_{EMO}^{h} = \begin{cases} p_{cs}^{h}E_{ex}^{h} - p_{cb}^{h}E_{im}^{h} + p_{gb}^{h}(E_{im}^{h} - E_{ex}^{h}), E_{im}^{h} < E_{ex}^{h} \\ p_{cs}^{h}E_{ex}^{h} - p_{cb}^{h}E_{im}^{h} + p_{gs}^{h}(E_{im}^{h} - E_{ex}^{h}), E_{im}^{h} > E_{ex}^{h} \end{cases}. \tag{9}$$

4. Optimization Scheduling

4.1. Rolling Optimization for Local MGs

The scheduling of the ESS is optimized based on the long term forecast information on PV generation, load demands and internal prices, which can be expressed as $pv_{i}^{h,f}$, $l_{i}^{h,f}$, $p_{cb}^{h,f}$ and $p_{cs}^{h,f}$, respectively. The primary target of a local MG is to maximize the operation utility. According to (2), the target of the rolling optimization is equivalent to minimize the cost of energy trading with the EMO, which can be expressed as

$$\min C_{i}^{roll} = \sum_{h}^{h+K\Delta h} [p_{cb}^{h,f} nl_{s,i}^{h} + p_{cs}^{h,f} nl_{b,i}^{h} + c_{i}e_{i}^{h2}], \tag{10}$$

where k is the length of rolling optimization, and Δh is the rolling step. In addition, the objective function in (10) needs to satisfy not only the constraints in (3)–(6) but also the constraints of battery charging and discharging, which can be expressed as follows.

$$e_{i}^{h} = e_{ch,i}^{h} - e_{dis,i}^{h}, \tag{11}$$

$$0 \leq e_{ch,i}^{h} \leq D_{ch,i}^{h} e_{ch,i}^{max}, \tag{12}$$

$$0 \leq e_{dis,i}^{h} \leq D_{dis,i}^{h} e_{dis,i}^{max}, \tag{13}$$

$$D_{ch,i}^{h} + D_{dis,i}^{h} \leq 1, \tag{14}$$

$$SOC_{i}^{h+1} = SOC_{i}^{h} + D_{ch,i}^{h} \cdot e_{ch,i}^{h} \cdot \eta_{ch,i} - D_{dis,i}^{h} \cdot e_{dis,i}^{h} / \eta_{dis,i}, \tag{15}$$

$$SOC_{i}^{min} \leq SOC_{i}^{h} \leq SOC_{i}^{max}, \tag{16}$$

where $e_{ch,i}^{h}$ and $e_{dis,i}^{h}$ are the charging and discharging energy of the ESS, respectively. $D_{ch,i}^{h}$ and $D_{dis,i}^{h}$ indicate the charging and discharging state of the ESS, $e_{ch,i}^{max}$ and $e_{dis,i}^{max}$ are the maximum charging and discharging energy in time slot h, respectively. Equation (15) shows the state of charge (SOC) of the ESS during its charging and discharging at the end of the time slot h, $\eta_{ch,i}$ and $\eta_{dis,i}$ are the charging and discharging efficiency of the ESS and SOC_{i}^{min} and SOC_{i}^{max} are the lower and upper limits of the SOC. Therefore, the rolling optimal scheduling in (10) can be modeled as a mixed integer programming (MIP) problem.

4.2. Stackelberg Game for EMOs

The optimal scheduling of the ESS in the first stage is regarded as one of the inputs of the EMO in second stage. The other inputs include the short term forecast information on PV generation, schedulable and unscheduled loads and the trading roles of the MG in the following hour. After receiving these inputs, the EMO will determine the internal prices and send to local EMSs for energy consumption optimization in the following hour.

4.2.1. Formulation of a Stackelberg Game

The hour-ahead energy sharing within the MMG is formulated as a Stackelberg game. The EMO is the leader of the game, and stimulates energy sharing by setting the internal prices with the goal of maximizing profits, while the MGs act as followers that optimize their utility through properly responding to internal prices. The game between the EMO and the MGs can be defined by its strategic form as

$$G = \left\{ (MGO \cup \{EMO\}), \left\{ L_i^h \right\}_{i \in N}, \left\{ P_{cb}^h \right\}, \left\{ P_{cs}^h \right\}, \left\{ U_i^h \right\}_{i \in N}, Pro_{MGC}^h \right\}, \tag{17}$$

where $\left\{ L_i^h \right\}_{i \in N}$ is the set of load strategies adopted by each MG i in the time slot h constrained by $l_i^{h,\min} \leq l_i^h \leq l_i^{h,\max}$; $\left\{ P_{cb}^h \right\}$ and $\left\{ P_{cs}^h \right\}$ are the strategic set of the EMO, which ensures that the internal prices are constrained by $p_{gs}^h \leq p_{cs}^h < p_{cb}^h \leq p_{gb}^h$; $\left\{ U_i^h \right\}_{i \in N}$ and Pro_{EMO}^h are the utility of the MG and the profit of the EMO which are expressed by (2) and (9), respectively.

Definition: Consider the game G defined in (17) as a set of strategies $(L_i^{h*}, P_{cb}^{h*}, P_{cs}^{h*})$ constituting a Stackelberg equilibrium (SE) if (and only if) the following set of inequalities are satisfied:

$$U_i^h(L_i^{h*}, P_{cb}^{h*}, P_{cs}^{h*}) \geq U_i^h(l_i^h, L_{-i}^{h*}, P_{cb}^{h*}, P_{cs}^{h*}) \ \forall i \in N, \forall l_i^h \in L_i^h, \tag{18}$$

$$Pro_{EMO}^h(L_i^{h*}, P_{cb}^{h*}, P_{cs}^{h*}) \geq Pro_{EMO}^h(L_i^{h*}, p_{cb}^h, p_{cs}^{h*}) \ \forall p_{cb}^h \in P_{cb}^h, \tag{19}$$

$$Pro_{EMO}^h(L_i^{h*}, P_{cb}^{h*}, P_{cs}^{h*}) \geq Pro_{EMO}^h(L_i^{h*}, p_{cb}^{h*}, p_{cs}^h) \ \forall p_{cs}^h \in P_{cs}^h, \tag{20}$$

where $L_i^{h*} = [l_1^{h*}, ..., l_i^{h*}, ..., l_N^{h*}]$, $L_{-i}^{h*} = [l_1^{h*}, ..., l_{i-1}^{h*}, l_{i+1}^{h*}, ..., l_N^{h*}]$. When all the players in $(MGO \cup \{EMO\})$ reach the SE, the EMO cannot improve its profit by adjusting the internal prices from the SE prices P_{cb}^{h*} and P_{cs}^{h*}. Likewise, no MGs can enhance their utilities by selecting different strategies from L_i^{h*}.

4.2.2. Achievement of Game Equilibrium

It is known from (2) and (3) that the utility of MGs can be modified as

$$U_i^h = \begin{cases} \theta p v_i^h + k_i^h \ln(1 + l_i^h) - p_{cb}^h(l_i^h + e_i^h - p v_i^h), & l_i^h + e_i^h - p v_i^h < 0 \\ \theta p v_i^h + k_i^h \ln(1 + l_i^h) - p_{cs}^h(l_i^h + e_i^h - p v_i^h), & l_i^h + e_i^h - p v_i^h \geq 0 \end{cases}. \tag{21}$$

For the given internal prices p_{cb}^h and p_{cs}^h, the optimal energy consumption l_i^{ho} can be easily obtained by making $\partial U_i^h / \partial l_i^h = 0$, which leads to

$$l_i^{ho} = \begin{cases} k_i^h / p_{cb}^h - 1, & l_i^h + e_i^h - p v_i^h < 0 \\ k_i^h / p_{cs}^h - 1, & l_i^h + e_i^h - p v_i^h > 0 \end{cases}. \tag{22}$$

Equation (22) can be substituted into (7)–(9) to solve the optimal internal price of the EMO based on the role of energy sharing that each MG wants to play. The following section provides a discussion of the relationship between the optimal prices and the roles of MGs.

For each MG acting as a buyer, the net energy nl_i^h satisfies $l_i^h + e_i^h - p v_i^h \geq 0$. The range of energy consumption can be redefined as

$$\max(l_i^{h,\min}, p v_i^h - e_i^h) \leq l_i^h \leq l_i^{h,\max}. \tag{23}$$

Substituting the optimal value in (22) into the energy consumption in (23), we can obtain

$$\frac{k_i^h}{l_i^{\max} + 1} \le p_{cs,i}^h \le \frac{k_i^h}{\max(l_i^{h,\min}, pv_i^h - e_i^h) + 1}, \tag{24}$$

where $p_{cs,i}^h$ is the flexible selling price that the MG i as a buyer expects to trade with the EMO. If $p_{cs}^h < k_i^h / (l_i^{\max} + 1)$, the optimal energy consumption will be $l_i^{\max}$; if $p_{cs,i}^h > k_i^h / (\max(l_i^{h,\min}, pv_i^h - e_i^h) + 1)$, the optimal energy consumption will be $\max(l_i^{h,\min}, pv_i^h - e_i^h)$. Similarly, the flexible buying price $p_{cb,i}^h$ of a seller is constrained by

$$\frac{k_i^h}{\min(pv_i^h - e_i^h, l_i^{h,\max}) + 1} \le p_{cb,i}^h \le \frac{k_i^h}{l_i^{\min} + 1}. \tag{25}$$

If $p_{cb}^h < k_i^h / (\min(pv_i^h - e_i^h, l_i^{h,\max}) + 1)$, the optimal energy consumption will be $\min(pv_i^h - e_i^h, l_i^{h,\max})$; if $p_{cs}^h > k_i^h / (l_i^{h,\min} + 1)$, the optimal energy consumption will be $\max(l_i^{h,\min}, pv_i^h - e_i^h)$. As a result, we can simplify (24) and (25) into

$$p_{cs,i}^h \in [p_{cs,i}^{h,\min}, p_{cs,i}^{h,\max}], \tag{26}$$

$$p_{cb,i}^h \in [p_{cb,i}^{h,\min}, p_{cb,i}^{h,\max}]. \tag{27}$$

Equation (26) expresses the feasible region of the optimal price for a seller and equation (27) expresses the feasible region of the optimal price for a buyer. Therefore, the profit functions of the EMO in (7)–(9) can be updated with the optimal energy consumption as

$$E_{ex}^{ho} = \sum_{p_{s,i}^{h,\min} \le p_{cs}^h \le p_{s,i}^{h,\max}} (\frac{k_i^h}{p_{cs}^h} - 1)D_{b,i}^h + \sum_{p_{cs}^h < p_{s,i}^{h,\min}} l_i^{h,\max}D_{b,i}^h + \sum_{p_{cs}^h > p_{s,i}^{h,\max}} \max(l_i^{h,\min}, pv_i^h - e_i^h)D_{b,i}^h + \sum_{D_{b,i}^h = 1} e_i^h - pv_i^h \tag{28}$$

$$E_{im}^{ho} = \sum_{p_{b,i}^{h,\min} \le p_{cb}^h \le p_{b,i}^{h,\max}} (\frac{k_i^h}{p_{cb}^h} - 1)D_{s,i}^h + \sum_{p_{cs}^h > p_{s,i}^{h,\max}} l_i^{h,\min}D_{s,i}^h + \sum_{p_{cb}^h < p_{b,i}^{h,\min}} \min(pv_i^h - e_i^h, l_i^{h,\max})D_{s,i}^h + \sum_{D_{s,i}^h = 1} e_i^h - pv_i^h \tag{29}$$

$$Pro_{MGC}^h = \begin{cases} p_{cs}^h E_{ex}^{ho} - p_{cb}^h E_{im}^{ho} + p_{gb}^h (E_{im}^{ho} - E_{ex}^{ho}), E_{im}^{ho} < E_{ex}^{ho} \\ p_{cs}^h E_{ex}^{ho} - p_{cb}^h E_{im}^{ho} + p_{gs}^h (E_{im}^{ho} - E_{ex}^{ho}), E_{im}^{ho} \ge E_{ex}^{ho} \end{cases}. \tag{30}$$

In order to optimize the objective function in (30), we define a two-dimensional coordinate system for the selling price and buying prices, as shown in Figure 3. The critical values in (26) of the sellers and the power grid prices are located in a transverse axis, such as $[p_{gb}^h, p_{cs,i}^{h,\min}, p_{cs,j}^{h,\min}, ..., p_{cs,k}^{h,\max}, p_{cs,l}^{h,\max}, p_{gs}^h]$, and the critical values in (27) of the buyers and the power grid prices are located in a longitudinal axis, such as $[p_{gb}^h, p_{cb,q}^{h,\min}, p_{cb,p}^{h,\min}, ..., p_{cs,m}^{h,\max}, p_{cs,n}^{h,\max}, p_{gs}^h]$. The feasible region S of the profit function (30) is divided into a certain number of sub-regions, such as sub-region $\forall s \in S$. For each sub-region $s \in S$, it is easily known that the profit function in (30) is concave (the proof can be seen in Appendix A). The optimal buying and selling prices in sub-region s can be found as $[p_{cb,s}^{h*}, p_{cs,s}^{h*}]$ by using CPLEX which is the commercial solver for MIP problem. Thus, the optimal prices in S can be obtained as

$$[p_{cb}^{h*}, p_{cs}^{h*}] = \arg\max_{(p_{cb}^h, p_{cs}^h)} \left(Pro_{MGC}^h (p_{cb,s}^{h*}, p_{cs,s}^{h*}), \forall s \in S \right). \tag{31}$$

Furthermore, the optimal prices $[p_{cb,s}^{h*}, p_{cs,s}^{h*}]$ will be sent to MGOs. The MGOs can calculate the optimal load demands L_i^{h*} according to formulas (22)–(27). Thus, the SE of the proposed game is reached. The sensitivity analysis of MGs' utilities over the internal prices adopted by the market operator are shown in Appendix B.

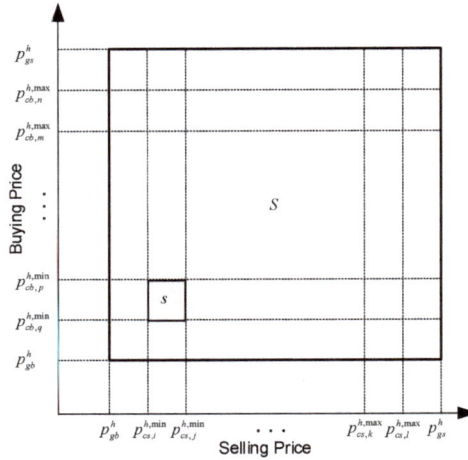

Figure 3. Feasible region of internal prices.

5. Case Studies

5.1. Basic Data

We employed MATLAB software to program the proposed model and analyze the simulation results. The MIP problem and convex optimization problem were solved by CPLEX. The model was applied to an MMG consisting of 3 MGs. All of the MGs had a PV system and an ESS installed, and the maximum schedulable loads were set to nearly 20% of the maximum load demand. The capacity of the ESS in each MG was 100kWh at the maximum charging/discharging rate of 0.5 C, and the range of SOC is from 0.2 to 1. The degradation cost coefficient c_i is 0.005.

The PV generation and load demands of the MGs in a typical day are shown in Figure 4, which were collected from the operation data of different MGs located in Jinzhai, Anhui Province, China. The time-of-use tariffs of the distribution network are shown in Table 1. The subsidy for PV energy was CNY 0.42 per kWh. The length of the rolling optimization was 8 hours at a step of 1 hour. The forecast information, which included load demands and PV generation, was generated from the collected historical data by using the prediction algorithms [20,21].

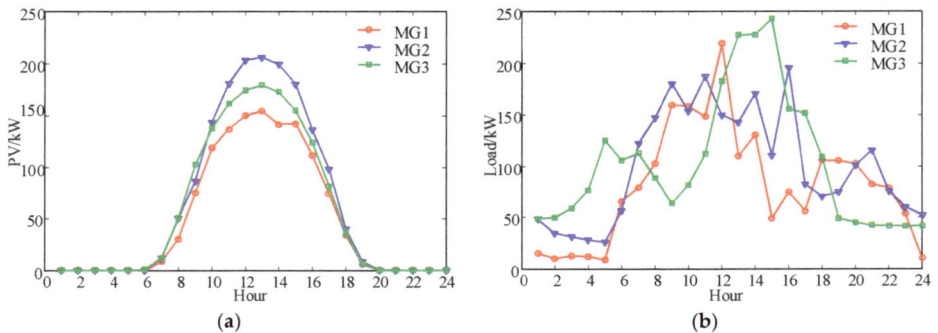

Figure 4. Basic data in hour: (**a**) PV energy output for 3 MGs, and (**b**) load demands for 3 MGs.

Table 1. Time-of-use tariff.

Distribution Network	Prices (kWh/h)	Hours
Selling	Peak: 1.189	8:00–11:00; 13:00–16:00; 18:00–22:00
	Flat: 0.738	7:00–8:00; 11:00–13:00; 16:00–18:00
	Valley: 0.423	0:00–7:00; 22:00–24:00
Buying	0.352	0:00–24:00

5.2. Internal Prices of the MMG

For the convenience of the following analysis, we provide the optimal prices in the MMG under the condition of market equilibrium as shown in Figure 5. It can be seen that that during the periods of 0:00–8:00 and 18:00–24:00 the internal buying and selling prices were equal to those of the power grid because the PV generation of each MG was very small, or even 0, and the internal load demand was relatively large. While all MGs are in the purchasing state, the EMO cannot improve its own benefits by adjusting the internal prices. During the period of 8:00–18:00, the internal buying price was always higher than that of the power grid, and the MG could sell more electricity by adjusting load demand to maximize operation utility. During the periods of 11:00–12:00 and 13:00–16:00, the internal sale price was lower than the selling price of the power grid, and the MG could increase energy consumption to reduce the purchasing cost of the MG and create more economic benefits.

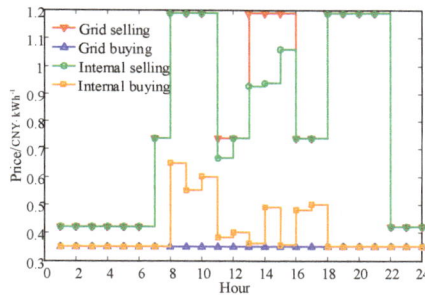

Figure 5. Comparison of internal prices and grid prices.

The above results show that the EMO could only adjust the internal electricity price during the period of 8:00–18:00, when PV generation is relatively strong. During the daytime, it could promote energy trading in the MMG, and realize the improvement of the operation efficiency of each MG. Next, we analyze the benefits of the MGs and the EMO in combination with the internal prices proposed in this paper.

5.3. Results of Local MGs

5.3.1. Rolling Optimization of ESSs

In the proposed method, the local ESS is charged in the case of a PV generation surplus or low electricity price, and discharged in the case of a PV energy deficit or high electricity price to reduce the cost at which the MG purchases the electricity from the power grid. Figure 6 shows the dispatching results of the ESSs in the local MGs. It can be seen that the ESS charged when the internal selling price was equal to the valley price of the power grid in the period of 0:00–6:00. In the period of 7:00–9:00, as load demands increased, the local PV energy was insufficient, the internal selling price was high and discharge of the ESS occurred to reduce the cost of buying energy from the power grids in that period. During the period of 15:00–18:00, the PV generation surplus was absorbed by the ESS. During the period of 18:00–22:00, when the electricity price was high, the absorbed energy was released to reduce the cost of power consumption in the MG.

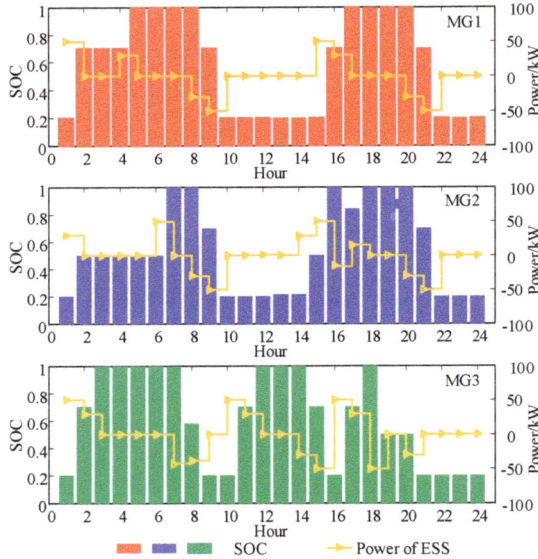

Figure 6. Optimal scheduling of energy storage systems (ESSs). SOC: state of charge.

In order to further analyze the benefits of ESS scheduling, the benefits of ESS charging and discharging in each period are shown in Figure 7. These benefits were calculated on the basis of the internal prices. The results show that the positive benefits of each ESS in a day were higher than the total negative benefits. The increased benefits of three MGs were CNY 130.08, CNY 133.66 and CNY 143.96, respectively. This validates the necessity and rationality of ESS scheduling in the first stage.

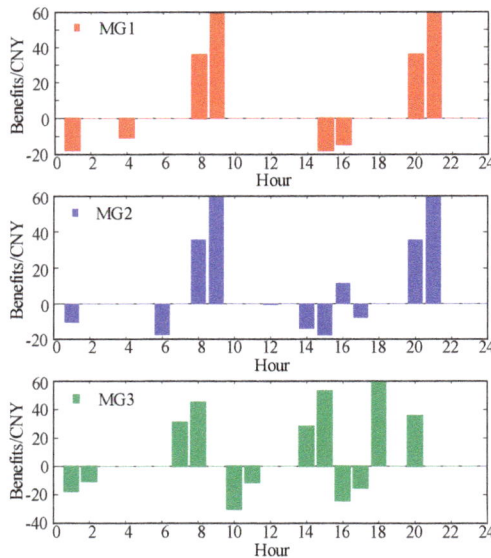

Figure 7. Impact of ESS scheduling on the benefits of MGs.

5.3.2. Demand Response of Local MGs

On the basis of optimal ESS scheduling, each MG will determine its own trading role (see Appendix C) to participate in energy trading in the MMGs by transferring the local information to the EMO. Under the incentive of internal prices, the MGs in the state of selling are encouraged to sell more energy to the EMO, and the MGs in the state of buying are encouraged to buy more energy from the EMO through the load adjustment as shown in Figure 8. During the periods of 11:00–12:00 and 13:00–16:00, MG1 and MG3 were stimulated by internal selling prices in the corresponding hours, which improved the operation utility by increasing energy consumption. On the contrary, the MG2 in the state of selling during the period of 7:00–18:00 could improve benefits by reducing load demand and selling more energy. Compared with the direct transaction with the power grid, the proposed method improves the utility of MGs, and the increased utility in each hour is shown in Figure 9.

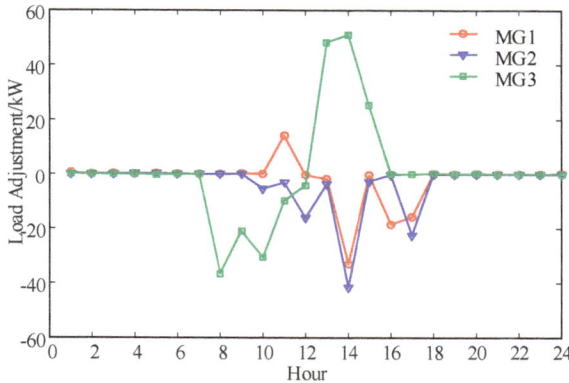

Figure 8. Adjustment of load demands.

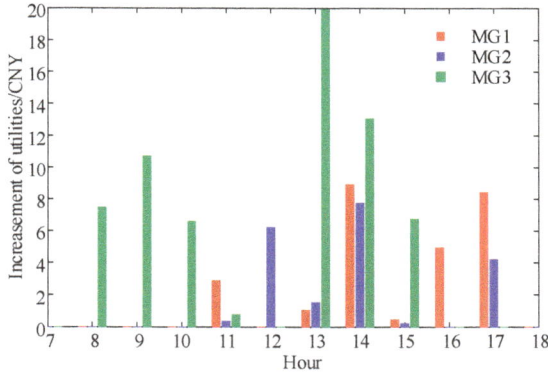

Figure 9. Increased utility of MGs based on internal prices.

5.4. Results of the EMO

The adjustment of energy consumption in local MGs directly affects the total energy sharing and total energy demand in the energy market, as shown in Figure 10. It can be seen that as the internal selling prices were lower than the grid selling prices during the periods of 11:00–12:00 and 13:00–16:00, the total energy demand in the energy market increased significantly. The total energy sharing in the energy market also increased as internal buying prices rose higher than the grid buying prices during the period of 7:00–18:00. The results show that MGs will actively participate in energy market transactions under the incentive of internal electricity prices.

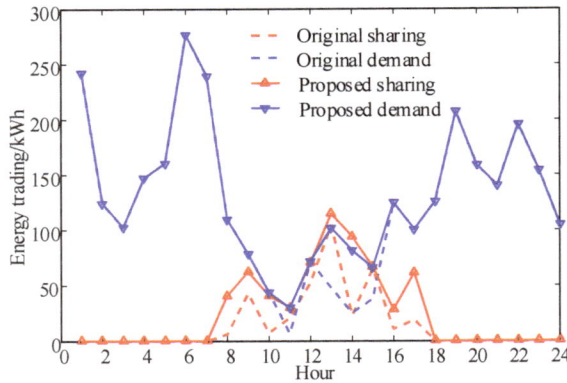

Figure 10. Comparison of energy trading in the energy market.

Energy demand and sharing in the energy market is the premise for the EMO to make profits. In the periods of 0:00–7:00 and 18:00–24:00, MGs were in the buying state thus the EMO could only purchase energy from the power grids to meet the demands of MGs and could not obtain profits. Figure 11 compares the profits of the EMO under the internal price strategy and the grid price strategy during the period of 7:00 and 18:00. Under the incentive of the internal price strategy, the profit of the EMO increased significantly in these periods. The daily profit of the EMO increased from CNY 171.6 to CNY 277.7 and the rate of increase is 61.82%.

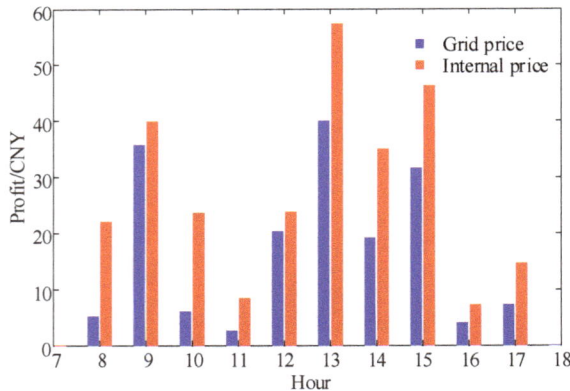

Figure 11. Comparison of the utility of the EMO.

To further illustrate the advantages of the proposed method, the net energy and PV utilization of the MMG were analyzed. The comparison of net energy curves is shown in Figure 12. During the period of 0:00–6:00, the charging of the ESS led to an increase in net energy. During the period of 18:00–24:00, the discharging of the ESS led to a decline in net energy. During the period of 6:00–18:00, the change in net energy was affected by ESS scheduling and the load demand response under internal prices. Obviously, the proposed method produced smaller fluctuations of the net load than the original method, and the PV energy reversal in the MMG during the period of 10:00–15:00 was significantly suppressed.

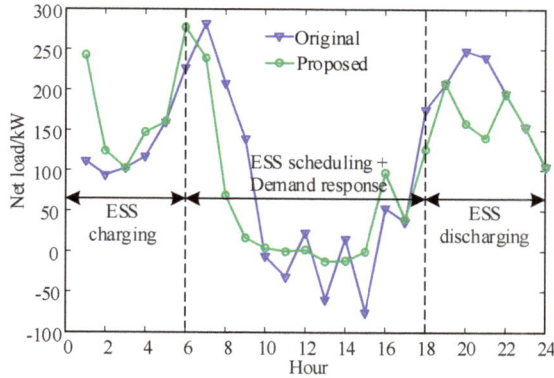

Figure 12. Comparison of net energy of MMGs.

Moreover, the comparison of peak-to-average ratio (PAR) and PV utilization ratio are shown in Table 2. Through comparisons, it is found that the proposed method can effectively reduce the peak-to-average ratio of the net load and improve the utilization ratio of PV energy.

Table 2. Comparison of pea-to-average ratio (PAR) and PV utilization ratio.

Method	PAR	PV Utilization Ratio
Original method	3.1596	85.25%
Proposed method	2.6992	98.07%

5.5. Utility Comparisons with Other Methods

To further illustrate the advantages of the proposed hour-ahead optimization over day-ahead optimization, the methods in references [16] and [17] were applied to the proposed case, and the cost optimization problem of references [16] and [17] were transformed into the utility optimization problem to make a comparative analysis of the results. In addition, the day-ahead stochastic prediction errors of PV and the load demand were set to 10% and 12%, while the hour-ahead stochastic prediction errors were set to 5% and 6%, respectively. The results are shown in Table 3.

Table 3. Comparisons with other methods.

Utility/CNY	Reference [16]	Reference [17]	Proposed
EMO	234.2	207.6	277.8
MG1	4792.3	4796.4	4823.0
MG2	5897.2	5952.0	5973.8
MG3	6897.8	6946.4	7011.6

The results show that EMO and MGO can achieve more benefits by using the optimization method proposed in this paper.

6. Conclusions

In this paper, a hierarchical optimization method is proposed to optimize the energy transaction of a MMG in two stages. Firstly, the EMS of each MG determines the scheduling of ESS in the next hour by adopting rolling optimization, and decides its participating role in energy trading market. Secondly, according to forecast information and the energy trading roles collected from MGs, the EMO optimizes the internal prices of next hour based on Stackelberg game theory. The simulation results show that the utility of both the EMO and MGO are increased by using the proposed method. In addition, the net load curve and utilization ratio of PV energy in the whole MMG system are both improved.

Author Contributions: Formal analysis, C.H., W.S. and B.X.; Funding acquisition, Q.W.; Methodology, T.R. and G.L.; Writing—original draft, T.R.

Funding: This work was supported by the National Key R&D Program of China (Nos. 2016YFB0900400)

Conflicts of Interest: The authors declare no conflict of interest.

Appendix A

It is assumed that

$$C_{ex} = \sum_{p_{cs}^h < p_{s,i}^{h,\min}} l_i^{h,\max} D_{b,i}^h + \sum_{p_{cs}^h > p_{s,i}^{h,\max}} \max(l_i^{h,\min}, pv_i^h - e_i^h) D_{b,i}^h + \sum_{D_{b,i}^h = 1} e_i^h - pv_i^h, \tag{32}$$

$$C_{im} = \sum_{p_{cs}^h > p_{s,i}^{h,\max}} l_i^{h,\min} D_{s,i}^h + \sum_{p_{cb}^h < p_{b,i}^{h,\min}} \min(pv_i^h - e_i^h, l_i^{h,\max}) D_{s,i}^h + \sum_{D_{s,i}^h = 1} e_i^h - pv_i^h. \tag{33}$$

Thus, expressions (32) and (33) can be simplified as

$$E_{ex}^{ho} = \sum_{p_{s,i}^{h,\min} \le p_{cs}^h \le p_{s,i}^{h,\max}} \left(\frac{k_i^h}{p_{cs}^h} - 1\right) + C_{ex}, \tag{34}$$

$$E_{im}^{ho} = \sum_{p_{b,i}^{h,\min} \le p_{cs}^h \le p_{b,i}^{h,\max}} \left(\frac{k_i^h}{p_{cb}^h} - 1\right) + C_{im}. \tag{35}$$

By substituting (42) and (43) into (34), the objective function of the EMO in sub-region $\forall s \in S$ is described as:

If $E_{im}^{ho} < E_{ex}^{ho}$

$$
\begin{aligned}
Pro_{EMO}^h &= p_{cs}^h \Big(\sum_{p_{s,i}^{h,\min} \le p_{cs}^h \le p_{s,i}^{h,\max}} \left(\frac{k_i^h}{p_{cs}^h} - 1\right) D_{b,i}^h + C_{ex} \Big) - p_{cb}^h \Big(\sum_{p_{b,i}^{h,\min} \le p_{cs}^h \le p_{b,i}^{h,\max}} \left(\frac{k_i^h}{p_{cb}^h} - 1\right) D_{s,i}^h + C_{im} \Big) \\
&+ p_{gb}^h \Big(\sum_{p_{b,i}^{h,\min} \le p_{cb}^h \le p_{b,i}^{h,\max}} \left(\frac{k_i^h}{p_{cb}^h} - 1\right) D_{s,i}^h + C_{im} \Big) - p_{gb}^h \Big(\sum_{p_{s,i}^{h,\min} \le p_{cs}^h \le p_{s,i}^{h,\max}} \left(\frac{k_i^h}{p_{cs}^h} - 1\right) D_{b,i}^h + C_{ex} \Big)
\end{aligned} \tag{36}
$$

If $E_{im}^{ho} \ge E_{ex}^{ho}$

$$
\begin{aligned}
Pro_{EMO}^h &= p_{cs}^h \Big(\sum_{p_{s,i}^{h,\min} \le p_{cs}^h \le p_{s,i}^{h,\max}} \left(\frac{k_i^h}{p_{cs}^h} - 1\right) D_{b,i}^h + C_{ex} \Big) - p_{cb}^h \Big(\sum_{p_{b,i}^{h,\min} \le p_{cs}^h \le p_{b,i}^{h,\max}} \left(\frac{k_i^h}{p_{cb}^h} - 1\right) D_{s,i}^h + C_{im} \Big) \\
&+ p_{gs}^h \Big(\sum_{p_{b,i}^{h,\min} \le p_{cs}^h \le p_{b,i}^{h,\max}} \left(\frac{k_i^h}{p_{cb}^h} - 1\right) D_{s,i}^h + C_{im} \Big) - p_{gs}^h \Big(\sum_{p_{s,i}^{h,\min} \le p_{cs}^h \le p_{s,i}^{h,\max}} \left(\frac{k_i^h}{p_{cs}^h} - 1\right) D_{b,i}^h + C_{ex} \Big)
\end{aligned} \tag{37}
$$

where $(p_{cb}^h, p_{cs}^h) \in s$, and $p_{cb}^h < p_{cs}^h$. Therefore, the Hessian matrix of Pro_{EMO}^h is

$$H = \begin{cases} \begin{bmatrix} -\dfrac{2p_{gb}^h}{p_{cs}^{h\,3}} \displaystyle\sum_{p_{s,i}^{h,\min} \leq p_{cs}^h \leq p_{s,i}^{h,\max}} D_{b,i}^h k_i^h & 0 \\[3em] 0 & \dfrac{2p_{gb}^h}{p_{cb}^{h\,3}} \displaystyle\sum_{p_{b,i}^{h,\min} \leq p_{cb}^h \leq p_{b,i}^{h,\max}} D_{s,i}^h k_i^h \end{bmatrix}, E_{im}^{ho} < E_{ex}^{ho} \\[6em] \begin{bmatrix} -\dfrac{2p_{gs}^h}{p_{cs}^{h\,3}} \displaystyle\sum_{p_{s,i}^{h,\min} \leq p_{cs}^h \leq p_{s,i}^{h,\max}} D_{b,i}^h k_i^h & 0 \\[3em] 0 & \dfrac{2p_{gs}^h}{p_{cb}^{h\,3}} \displaystyle\sum_{p_{b,i}^{h,\min} \leq p_{cb}^h \leq p_{b,i}^{h,\max}} D_{s,i}^h k_i^h \end{bmatrix}, E_{im}^{ho} \geq E_{ex}^{ho} \end{cases} \quad (38)$$

Consider the fact that $k_i^h > 0$, $p_{gb}^h > 0$, $p_{gs}^h > 0$, $p_{cs}^h > 0$, $p_{cb}^h > 0 \ \forall i \in N, h \in H$, H is thus negative definite and Pro_{MGC}^h is strictly concave with respect to p_{cs}^h and p_{cb}^h.

Appendix B

Formulas (26) and (27) show that if $p_{cs,i}^h \notin [p_{cs,i}^{h,\min}, p_{cs,i}^{h,\max}]$, then $\partial U_i^h / \partial p_{cs,i}^h = 0$, and if $p_{cb,i}^h \notin [p_{cb,i}^{h,\min}, p_{cb,i}^{h,\max}]$, then $\partial U_i^h / \partial p_{cb,i}^h = 0$. Therefore, we mainly discuss the case of $p_{cs,i}^h \in [p_{cs,i}^{h,\min}, p_{cs,i}^{h,\max}]$ and $p_{cb,i}^h \in [p_{cb,i}^{h,\min}, p_{cb,i}^{h,\max}]$. For MG i, who is a buyer, if $p_{cs,i}^h \in [p_{cs,i}^{h,\min}, p_{cs,i}^{h,\max}]$, then the optimal load demand can be expressed as

$$l_i^{ho} = k_i^h / p_{cb}^h - 1 = (l_i^{hf} + 1)p_{gb}^h / p_{cb}^h - 1. \quad (39)$$

The corresponding utility is rewritten as

$$U_i^h = \theta p v_i^h + (l_i^{hf} + 1)p_{gb}^h \ln\left(\frac{(l_i^{hf} + 1)p_{gb}^h}{p_{cb}^h}\right) - p_{cb}^h\left(\frac{(l_i^{hf} + 1)p_{gb}^h}{p_{cb}^h} - 1 + e_i^h - p v_i^h\right). \quad (40)$$

Therefore, the derivative of benefit U_i^h with respect to price p_{cb}^h is

$$\frac{\partial U_i^h}{\partial p_{cb,i}^h} = \frac{(l_i^{hf} + 1)p_{gb}^h}{p_{cb}^h} - 1 + e_i^h - p v_i^h. \quad (41)$$

For MG i, who is a seller, if $p_{cb,i}^h \in [p_{cb,i}^{h,\min}, p_{cb,i}^{h,\max}]$, then the optimal load demand can be expressed as

$$l_i^{ho} = k_i^h / p_{cs}^h - 1 = (l_i^{hf} + 1)p_{gb}^h / p_{cs}^h - 1. \quad (42)$$

The corresponding utility is rewritten as

$$U_i^h = \theta p v_i^h + (l_i^{hf} + 1)p_{gb}^h \ln\left(\frac{(l_i^{hf} + 1)p_{gb}^h}{p_{cs}^h}\right) - p_{cs}^h\left(\frac{(l_i^{hf} + 1)p_{gb}^h}{p_{cs}^h} - 1 + e_i^h - p v_i^h\right). \quad (43)$$

Therefore, the derivative of benefit U_i^h with respect to price p_{cs}^h is

$$\frac{\partial U_i^h}{\partial p_{cs,i}^h} = \frac{(l_i^{hf} + 1)p_{gb}^h}{p_{cs}^h} - 1 + e_i^h - p v_i^h. \quad (44)$$

To further demonstrate the utility of MG response to the internal electricity price, the utilities of MG1 and MG3 in time slot 14 were chosen to show the relationship between utility and changing prices, which are shown in Figures A1 and A2, respectively.

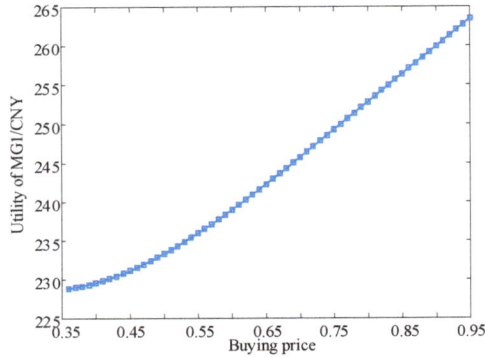

Figure A1. The utility of MG1 response to internal buying prices. ($k_1^{14} = 46.1$, $l_1^{14f} = 130.7$, $pv_1^{14} = 141.6$, $e_1^{14} = 0$, $p_{cs}^{14} = 0.95$, $p_{gb}^{14} = 0.35$, $p_{gs}^{14} = 1.189$).

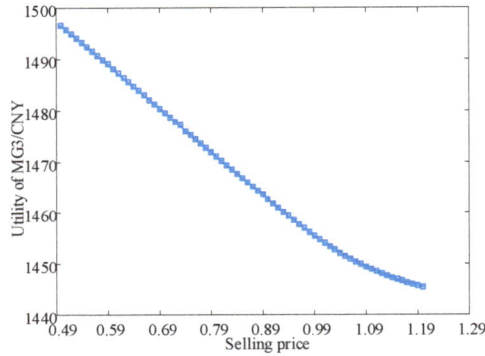

Figure A2. The utility of MG3 response to internal selling prices. ($k_3^{14} = 271.6$, $l_3^{14f} = 227.4$, $pv_3^{14} = 172.7$, $e_3^{14} = -30$, $p_{cb}^{14} = 0.49$, $p_{gb}^{14} = 0.35$, $p_{gs}^{14} = 1.189$).

Appendix C

The optimal results of $D_{b,i}^h$ and $D_{b,i}^h$ determine the roles of MGs participating in energy trading, which are all listed in the follow Table A1.

Table A1. Roles of Energy Trading.

Period	MG1	MG2	>MG3
0:00–8:00	buyer	buyer	buyer
8:00–9:00	buyer	buyer	seller
9:00–10:00	buyer	buyer	seller
10:00–11:00	buyer	buyer	seller
11:00–12:00	buyer	buyer	seller
12:00–13:00	buyer	seller	buyer
13:00–14:00	seller	seller	buyer
14:00–15:00	seller	seller	buyer
15:00–16:00	seller	seller	buyer
16:00–17:00	seller	buyer	buyer
17:00–18:00	seller	seller	buyer
18:00–24:00	buyer	buyer	buyer

References

1. Rafique, S.F.; Zhang, J. Energy management system, generation and demand predictors: A review. *IET Gener. Transm. Distrib.* **2018**, *7*, 519–530. [CrossRef]
2. Han, Y.; Zhang, K.; Li, H.; Coelho, E.A.; Guerrero, J.M. MAS-based Distributed Coordinated Control and Optimization in Microgrid and Microgrid Clusters: A Comprehensive Overview. *IEEE Trans. Power Electron.* **2018**, *33*, 6488–6508. [CrossRef]
3. Nguyen, T.A.; Crow, M.L. Stochastic Optimization of Renewable-Based Microgrid Operation Incorporating Battery Operating Cost. *IEEE Trans. Power Syst.* **2016**, *31*, 2289–2296. [CrossRef]
4. Liu, W.; Gu, W.; Sheng, W.; Meng, X.; Wu, Z.; Chen, W. Decentralized Multi-Agent System-Based Cooperative Frequency Control for Autonomous Microgrids With Communication Constraints. *IEEE Trans. Sustain. Energy* **2014**, *5*, 446–456. [CrossRef]
5. Li, Q.; Chen, F.; Chen, M.; Guerrero, J.M.; Abbott, D. Agent-Based Decentralized Control Method for Islanded Microgrids. *IEEE Trans. Smart Grid* **2016**, *7*, 637–649. [CrossRef]
6. Parisio, A.; Wiezorek, C.; Kyntäjä, T.; Elo, J.; Strunz, K.; Johansson, K.H. Cooperative MPC-Based Energy Management for Networked Microgrids. *IEEE Trans. Smart Grid* **2017**, *8*, 3066–3074. [CrossRef]
7. Najmeh, B.; Ahmadreza, T.; Amjad, A.; Josep, M.G. Optimal operation management of a regionalnetwork of microgrids based on chanceconstrained model predictive control. *IET Gener. Transm. Distrib.* **2018**, *12*, 3772–3779.
8. Rafiee Sandgani, M.; Sirouspour, S. Priority-based Microgrid Energy Management in a Network Environment. *IEEE Trans. Sustain. Energy* **2018**, *9*, 980–990. [CrossRef]
9. Jadhav, A.M.; Patne, N.R.; Guerrero, J.M. A Novel Approach to Neighborhood Fair Energy Trading in a Distribution Network of Multiple Microgrid Clusters. *IEEE Trans. Ind. Electron.* **2018**, *66*, 1520–1531. [CrossRef]
10. Zhang, B.; Li, Q.; Wang, L.; Feng, W. Robust optimization for energy transactions in multi-microgrids under uncertainty. *Appl. Energy* **2018**, *217*, 346–360. [CrossRef]
11. Bui, V.H.; Hussain, A.; Kim, H.M. A Multiagent-Based Hierarchical Energy Management Strategy for Multi-Microgrids Considering Adjustable Power and Demand Response. *IEEE Trans. Smart Grid* **2018**, *9*, 1323–1333. [CrossRef]
12. Maryam, M.; Hassan, M.; Amjad, A.; Josep, M.G.; Hamid, L. A decentralized robust model for optimal operation of distribution companies with private microgrids. *Electr. Power Energy Syst.* **2019**, *106*, 105–123.
13. Manshadi, S.D.; Khodayar, M.E. A Hierarchical Electricity Market Structure for the Smart Grid Paradigm. *IEEE Trans. Smart Grid* **2016**, *7*, 1866–1875. [CrossRef]
14. Yue, J.; Hu, Z.; Amjad, A.M.; Josep, M.G. A Multi-Market-Driven Approach to Energy Scheduling of Smart Microgrids in Distribution Networks. *Sustainability* **2019**, *11*, 301. [CrossRef]
15. Fan, S.; Ai, Q.; Piao, L. Bargaining-based cooperative energy trading for distribution company and demand response. *Appl. Energy* **2018**, *226*, 469–482. [CrossRef]
16. Liu, Y.; Guo, L.; Wang, C. A robust operation-based scheduling optimization for smart distribution networks with multi-microgrids. *Appl. Energy* **2018**, *228*, 130–140. [CrossRef]
17. Jalali, M.; Zare, K.; Seyedi, H. Strategic decision-making of distribution network operator with multi-microgrids considering demand response program. *Energy* **2017**, *141*, 1059–1071. [CrossRef]
18. Liu, N.; Cheng, M.; Yu, X.; Zhong, J.; Lei, J. Energy Sharing Provider for PV Prosumer Clusters: A Hybrid Approach using Stochastic Programming and Stackelberg Game. *IEEE Trans. Ind. Electron.* **2018**, *65*, 6740–6750. [CrossRef]
19. Liu, N.; Yu, X.; Wang, C.; Wang, J. Energy Sharing Management for Microgrids with PV Prosumers: A Stackelberg Game Approach. *IEEE Trans. Ind. Inform.* **2017**, *13*, 1088–1098. [CrossRef]
20. Abedinia, O.; Amjady, N.; Zareipour, H. A New Feature Selection Technique for Load and Price Forecast of Electrical Power Systems. *IEEE Trans. Power Syst.* **2017**, *32*, 62–74. [CrossRef]
21. Gigoni, L.; Betti, A.; Crisostomi, E. Day-Ahead Hourly Forecasting of Power Generation from Photovoltaic Plants. *IEEE Trans. Sustain. Energy* **2018**, *9*, 831–842. [CrossRef]

![applied sciences logo] *applied sciences*

MDPI

Article

Optimal Non-Integer Sliding Mode Control for Frequency Regulation in Stand-Alone Modern Power Grids

Zahra Esfahani [1], Majid Roohi [2], Meysam Gheisarnejad [3], Tomislav Dragičević [4] and Mohammad-Hassan Khooban [5,*]

[1] Department of Electrical Engineering, University College of Rouzbahan, Sary 3994548179, Iran
[2] School of Economics and Statistics, Guangzhou University, Guangzhou 510006, China
[3] Department of Electrical Engineering, Najafabad Branch, Islamic Azad University, Isfahan 8514143131, Iran
[4] Department of Energy Technology, Aalborg University, DK-9220 Aalborg East, Denmark
[5] Department of Engineering, Aarhus University, 8200 Aarhus N, Denmark
[*] Correspondence: khooban@eng.au.dk

Received: 28 June 2019; Accepted: 9 August 2019; Published: 19 August 2019

Abstract: In this paper, the concept of fractional calculus (FC) is introduced into the sliding mode control (SMC), named fractional order SMC (FOSMC), for the load frequency control (LFC) of an islanded microgrid (MG). The studied MG is constructed from different autonomous generation components such as diesel engines, renewable sources, and storage devices, which are optimally planned to benefit customers. The coefficients embedded in the FOSMC structure play a vital role in the quality of controller commands, so there is a need for a powerful heuristic methodology in the LFC study to adjust the design coefficients in such a way that better transient output may be achieved for resistance to renewable sources fluctuations. Accordingly, the Sine Cosine algorithm (SCA) is effectively combined with the harmony search (HS) for the optimal setting of the controller coefficients. The Lyapunov function based on the FOSMC is formulated to guarantee the stability of the LFC mechanism for the test MG. Finally, the hardware-in-the-loop (HIL) experiments are carried out to ensure that the suggested controller can suppress the frequency fluctuations effectively, and that it provides more robust MG responses in comparison with the prior art techniques.

Keywords: non-integer sliding mode control; modified sine–cos optimization algorithm; islanded microgrid; nonlinear robust control

1. Introduction

Over the past decade, the reduction of conventional fossil fuel reserves, along with the environmental concerns about their burning, have led to the paradigm change toward the development of renewable energy sources (RESs), such as photovoltaic (PV) and wind turbine generator (WTG) systems penetrating into the power grid [1–3]. The planning and exploitation of RESs through distributed generators (DGs) offers the benefits of local generation according to the needs of consumers, with a consequent minimization in the transmission loss. The DGs can be potentially implemented for improving power quality and service reliability [4–8].

A microgrid is regarded as a regulated entity in the power plant and consists of various DGs such as microsources, energy reserve devices, and loads which are locally integrated into the grid for the profit of the customers. Normally, microgrids (MGs) can work in the stand-alone mode as independent islands or grid-connected modes in conjunction with the main electrical grid [9–11]. In the grid-connected mode, the main utility generators are responsible for power balance management in connection with new demands. In the isolated operation, the DGs are exposed to the load variations and randomness characteristics of renewable sources, which threaten the grid's stability. The inherent weakness of the integrated power plants may be somewhat shorted out by the installation of backup systems such as microturbines and energy storage systems (ESSs) [12,13].

Since load frequency control (LFC) capacity is not sufficient to meet all of the control objectives of the microgrid, employing a supplementary regulator in a secondary loop is obligatory. To deal with the secondary LFC control of an islanded MG, the conventional proportional integral derivative (PID) controllers and their variants are capable of restoring frequency deviations in normal operation; however, they cannot work appropriately as operation points of the grid vary remarkably during a daily cycle. To improve the efficiency of the LFC, some suitable control strategies including H∞ control theory [7], adaptive control [14], robust control [15], and model predictive control (MPC) [16] have been applied to the DGs of the hybrid MG. In this regard, an MPC-based coordinated control of the blade pitch angles of the wind turbine (WT), and the plug-in hybrid electric vehicle (PHEV), has been developed in [16] for the LFC issue. The authors in [17] proposed a robust controller using linear matrix inequalities to control the frequency in an integrated form of the MG and multi-MG with different DG units. A model-free intelligent proportional integral (PI) controller is suggested in [18] for several configurations of the isolated hybrid RESs/ESSs. The influence of the increased penetration of WTs and MGs on frequency control is investigated in [19], and an advanced hierarchical control methodology is established for the optimal control of the whole system.

The sliding mode control (SMC) is recognized as a prominent model-based approach to handle disturbances, since this design scheme has inherent insensitivity features against the dynamical system uncertainty. The advantages of the SMC are ease of use, quick convergent velocity, and robustness. Up to now, several published works have been done to boost the strength of the SMC by combining it with H∞ control theory [20], fuzzy logic [21] and neural network [22], etc. However, implementing such sophisticated hybrid methodologies is a challenging duty for controller designers. Most recently, similar to the uses of the fractional calculus (FC) concepts in other conventional methodologies [23,24], the fractional order is incorporated for extension of the simple classic version of the SMC, so that the degree of flexibility is enhanced. A few types of studies have been published for the application of the fractional order SMC (FOSMC) in engineering applications [25,26].

In this study, a cooperative combination of the fractional order and SMC scheme is developed and implemented for the grid frequency control in a hybrid MG. The simulation study is accomplished on a complex MG, including various RESs, to indicate the significance of the suggested model-based FOSMC control theory. The classic methodologies, as mentioned earlier, may not be applicable in such circumstances to guarantee the stability of the hybrid power plant. The controller parameters are adjusted automatically by using an effective combination of the Sine Cosine algorithm and harmony search (SCA-HS). Designing the suggested control strategy is based on the model specifications of the components configured in the controlled MG, and thus the time domain design of such a scheme is very valuable. The proposed FOSMC scheme offers superior frequency regulation of MGs, which are composed of numerous DGs and RESs, in comparison with the MPC and conventional SMC approaches. Moreover, the proposed scheme has more robustness, to tackle more uncertainties than the above-mentioned approaches, making it more suitable for practical applications. For investigating the performance and robustness of the suggested model-based FOSMC technique, experimental validation using hardware-in-the-loop (HIL) simulations are also given in this paper.

2. Non-Integer Order Calculus

Definition 1. *The Riemann–Liouville fractional integration of order α of a continuous function $f(t)$ is defined by [27]*

$$D^{-\alpha}f(t) = \frac{1}{\Gamma(\alpha)} \int_{t_0}^{t} f(\tau)(t-\tau)^{\alpha-1}d\tau \tag{1}$$

where t_0 is the initial time and $\Gamma(.)$ is the Gamma function, which is defined by $\Gamma(z) = \int_{t_0}^{\infty} t^{z-1}e^{-t}dt$.

Definition 2. *Let* $m - 1 < \alpha \leq m$ *and* $m \in N$, *then the Caputo fractional derivative of order* α *of a continuous function* $\varphi(t) : R^+ \to R$ *is given by* [27]:

$$D^{\alpha}\varphi(t) = \frac{1}{\Gamma(m-\alpha)} \int_{t_0}^{t} \frac{\varphi^{(m)}(\tau)}{(t-\tau)^{\alpha-m+1}} d\tau \tag{2}$$

Property 1 ([28]). *For any constant,* $\xi \in R$ *we have* $D^{\alpha}\xi = 0$.

Property 2 ([28]). *For* $\alpha \in (0,1)$ *and* $\omega(t) \in C^m[0,T]$, *we have:*

$$D^{\alpha}(I^{\alpha}\omega(t)) = D^{\alpha}(D^{-\alpha}\omega(t)) = \omega(t) \tag{3}$$

Lemma 1. *Suppose that* $\psi(t) \in C^m[0,T]$ *and* $\alpha \in (0,1)$. *Then,*

$$D^{\alpha}|\psi(t)| = sgn(\psi(t))D^{\alpha}\psi(t) \tag{4}$$

Proof. As $\alpha \in (0,1)$, because of Equation (3), one gets:

$$D^{\alpha}|\psi(t)| = \frac{1}{\Gamma(1-\alpha)} \int_{t_0}^{t} \frac{|\psi(t)|'}{(t-\tau)^{\alpha}} d\tau$$

Besides, $|\psi(t)|' = sgn(\psi(t)).\psi(t)'$, so in Equation (2), one has:

$$D^{\alpha}|\psi(t)| = \frac{1}{\Gamma(1-\alpha)} \int_{t_0}^{t} \frac{sgn(\psi(t))\psi(t)'}{(t-\tau)^{\alpha}} d\tau \tag{5}$$

□

Theorem 1 ([29]). *Suppose that for* $\alpha \in (0, 1)$, *the fractional-order system* $D^{\alpha}y(t) = g(y, t)$ *satisfies the Lipschitz condition and it has an equilibrium point like* $y = 0$. *Assume that there exists a Lyapunov function* $V(t, y(t))$ *and class-K functions* α_1, α_2 *and* α_3 *satisfying*

$$\alpha_1(\|y\|) \leq V(t, y) \leq \alpha_2(\|y\|) \tag{6}$$

$$D^p V(t, y) \leq -\alpha_3(\|y\|) \tag{7}$$

which $p \in (0,1)$. *Then the equilibrium point of the system* $D^{\alpha}y(t) = g(y, t)$ *is asymptotically stable.*

Theorem 2 ([30]). *Consider the following FO system*

$$D^{\alpha}X = F(x, t). \tag{8}$$

Let $\Lambda : (0, \infty) \times [0, X] \to R^n$ *introduce as*

$$\Lambda(\omega, t) = \int_{0}^{t} e^{-\omega^2(t-\theta)} F(x, \theta) d(\theta) \tag{9}$$

Then, the FO system (8) can be written as

$$\begin{cases} \frac{\partial \Lambda(\omega,t)}{\partial t} = -\omega^2 \Lambda(\omega, t) + F(x, t) \\ X(t) = \int_{0}^{\infty} u(\omega)\Lambda(\omega, t)d\omega \end{cases} \tag{10}$$

where $u(\omega) = \frac{2\sin(\alpha\pi)}{\pi}\omega^{1-2\alpha}$, $\alpha \in (0,1)$.

3. Description of an Isolated Fractional-Order Microgrid Model

3.1. An Isolated Microgrid

In an isolated MG, the distributed loads are provided by various DG components such as PVs and WTGs, and backup system elements (e.g., battery energy storage system (BESS) and flywheel energy storage system (FESS)) [8,31]. A general scheme of microgrids is illustrated in Figure 1. Usually, the MG dispatch system (MGDS) and the distribution management system (DMS) control the MG operation and the power grid, respectively. Moreover, communication links attain reciprocal information transition.

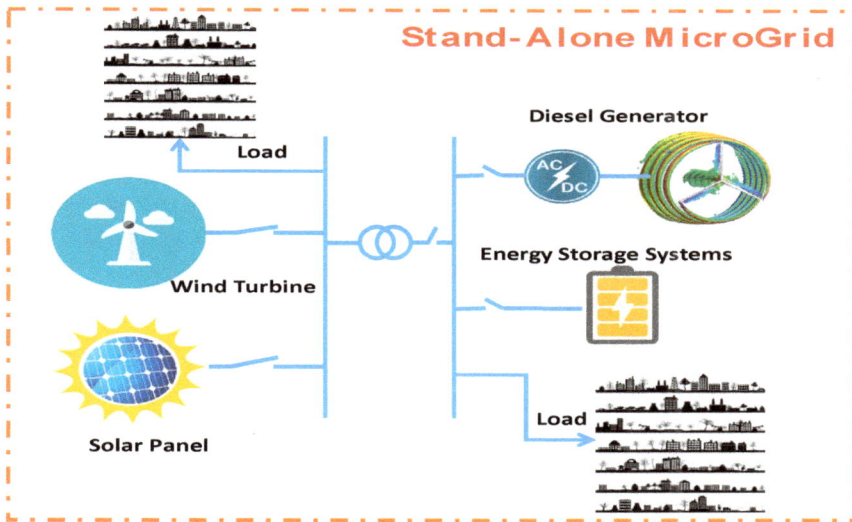

Figure 1. A general configuration of microgrids.

3.2. The Diesel Engine Generator Model

The diesel engine generators (DEGs) have a lot of advantages, such as their fast speed in the start, low maintenance, and high efficiency; hence, they have been a good option for backup in isolated MGs [32]. By precisely regulating the DEG, the changes of load in the MGs can be instantaneously tracked. In addition, the diesel power element can compensate for the fluctuations of renewable DGs (such as WTG, PV, etc.) effectively.

The transfer function of the DEG is illustrated in Figure 2, which describes the control relationship of the DEG output power and the LFC action. As presented in Figure 2, the elements of the governor and generator are represented by the first-order inertia models of inertia term.

3.3. Wind Turbine Generator

The output power of wind turbines depends on the inherent specifications of the turbine, and two factors: The speed and direction of the wind. The controllable WTG can be considered as a power oscillation source in the MG via the control of the DG sections.

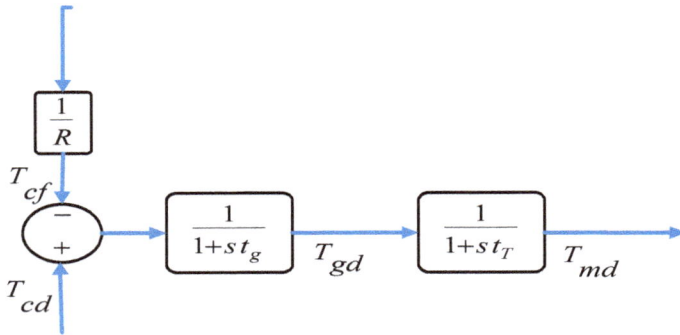

Figure 2. Model of the diesel engine generator (DEG) system.

3.4. Model of a Photovoltaic (PV) Generation

The electrical energy can be produced from the energy of the photons in PV cells, which are constructed from semiconductor materials. Due to the external and boundary contiguity along with the series resistance, losing power is inevitable in PVs. Naturally, the PVs have an intermittent characteristic, and their generated power depends on several factors, such as the radiation intensity, the surface area of the cell, and the ambient temperature [11]. A random power source can be utilized to model the stochastic behavior of PVs in simulations.

3.5. Structure of the LFC-Based MG System

Figure 3 illustrates the structure of the suggested LFC for the test MG, which employs different DGs such as PV, WT and DEG, storage devices (BESS and FESS), and loads. It is clear that the PV, fuel cell (FC), and BESS parts are connected to the AC MG via DC/AC interfacing inverters. All small-scale DGs and energy storage sections are connected to the AC bus via a circuit breaker. The spinning reserve for the secondary frequency control is offered by the diesel power system.

The dynamics of the MG system is shown in Figure 3. This system is a nine-state set of fractional-order equations. The objects of state equations are presented in the following:

$$
\begin{aligned}
D^\beta(\Delta f) &= \tfrac{1}{2H}\Big[T_{s_{filt}} + T_{md} + T_w - \Delta P_L + T_{f_{filt}} - T_{bat} - D * \Delta f\Big] \\
D^\beta(T_{s_{inv}}) &= \tfrac{1}{T_{inv}}\Big[T_s - T_{s_{inv}}\Big] \\
D^\beta(T_{s_{filt}}) &= \tfrac{1}{T_{filt}}\Big[T_{s_{inv}} - T_{s_{filt}}\Big] \\
D^\beta(T_{gd}) &= \tfrac{1}{T_g}\Big[T_{cd} - \tfrac{\Delta f}{R} - T_{gd}\Big] \\
D^\beta(T_{md}) &= \tfrac{1}{T_t}\Big[T_{gd} - T_{md}\Big] \\
D^\beta(T_{fc}) &= \tfrac{1}{T_{fc}}\Big[T_{cf} - \tfrac{\Delta f}{R} - T_{fc}\Big] \\
D^\beta(T_{f_{inv}}) &= \tfrac{1}{T_{inv}}\Big[T_{fc} - T_{f_{inv}}\Big] \\
D^\beta(T_{f_{filt}}) &= \tfrac{1}{T_{filt}}\Big[T_{f_{inv}} - T_{f_{filt}}\Big] \\
D^\beta(T_{bat}) &= \tfrac{1}{T_b}\Big[\Delta f - T_{bat}\Big]
\end{aligned}
\tag{11}
$$

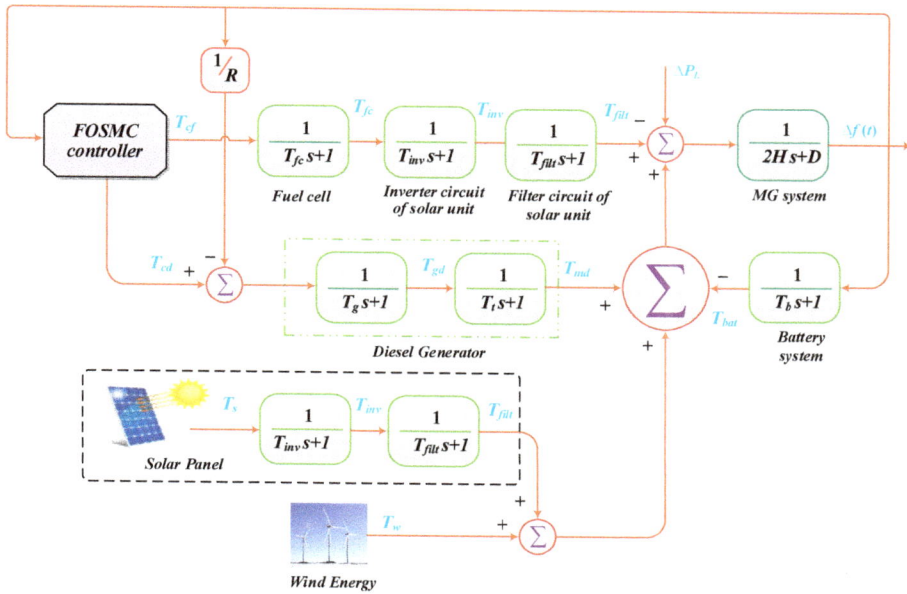

Figure 3. The overall microgrid (MG) scheme for load frequency control (LFC). FOSMC: fractional order sliding mode control.

The nine state MG dynamics and the output equation rewrite in a compact form in the following equations:

$$D^\alpha X = AX + BU$$
$$Y = CX + DU \tag{12}$$

where X, A, B, C, D, and U are stated matrix, system matrix, input matrix, the direct transition matrix, and input respectively, and they are introduced as follows:

$$X = \begin{bmatrix} \Delta f\ T_{s_{inv}}\ T_{s_{filt}}\ T_{gd}\ T_{md}\ T_{fc}\ T_{f_{inv}}\ T_{f_{filt}}\ T_{bat} \end{bmatrix}^T$$

$$A = \begin{bmatrix}
-\frac{D}{2H} & 0 & \frac{1}{2H} & 0 & \frac{1}{2H} & 0 & 0 & \frac{1}{2H} & -\frac{1}{2H} \\
0 & -\frac{1}{t_{inv}} & 0 & 0 & 0 & 0 & 0 & 0 & 0 \\
0 & \frac{1}{t_{filt}} & -\frac{1}{t_{filt}} & 0 & 0 & 0 & 0 & 0 & 0 \\
-\frac{1}{Rt_g} & 0 & 0 & \frac{1}{t_g} & 0 & 0 & 0 & 0 & 0 \\
0 & 0 & 0 & \frac{1}{t_T} & -\frac{1}{t_T} & 0 & 0 & 0 & 0 \\
\frac{1}{Rt_{fc}} & 0 & 0 & 0 & 0 & -\frac{1}{t_{fc}} & 0 & 0 & 0 \\
0 & 0 & 0 & 0 & 0 & \frac{1}{t_{inv}} & -\frac{1}{t_{inv}} & 0 & 0 \\
0 & 0 & 0 & 0 & 0 & 0 & \frac{1}{t_{filt}} & -\frac{1}{t_{filt}} & 0 \\
\frac{1}{t_b} & 0 & 0 & 0 & 0 & 0 & 0 & 0 & -\frac{1}{2H}
\end{bmatrix}$$

$$
B = \begin{bmatrix} 0 & 0 \\ 0 & 0 \\ 0 & 0 \\ 1/t_g & 0 \\ 0 & 0 \\ 0 & 1/t_{fc} \\ 0 & 0 \\ 0 & 0 \\ 0 & 0 \end{bmatrix}, N = \begin{bmatrix} -1/2H & 1/2H & 0 \\ 0 & 0 & 1/t_{inv} \\ 0 & 0 & 0 \\ 0 & 0 & 0 \\ 0 & 0 & 0 \\ 0 & 0 & 0 \\ 0 & 0 & 0 \\ 0 & 0 & 0 \\ 0 & 0 & 0 \end{bmatrix},
$$

$$
U = \begin{bmatrix} T_{cd} \; T_{cf} \end{bmatrix}^T, Q = [\Delta P_L \; T_w \; T_s]^T
$$
$$
C = [1\,0\,0\,0\,0\,0\,0\,0\,0], D = 0.
$$

Table 1 describes the parameters of the MG equations utilized in the simulation.

Table 1. The MG power system's parameters.

Symbol and Abbreviation	Values	Symbol and Abbreviation	Values
ΔP_L (change in load power)	0.02 s	T_g (governor time constant)	0.08 s
T_{inv} (time constant of inverter circuit of solar unit)	0.04 s	T_b (battery power time constant)	0.1 s
T_{fc} (time constant of fuel cell)	0.26 s	2H (inertia constant)	0.1667
T_{filt} (time constant of filter circuit of solar unit)	0.004 s	D (damping coefficient)	0.015
T_t (diesel generator time constant)	2.00 s	R (DG speed regulation)	3

4. Proposed Fractional-Order Sliding Mode Control Scheme

Here, a non-integer sliding surface is designed as follows:

$$
S(t) = X + D^{-\alpha}\big[K(|X|^p)sign(X) \big] \tag{13}
$$

where $1 < p < 2$ and K, is a vector of positive constants.

Based on the sliding mode control theory, when the system operates in the sliding mode, it should satisfy:

$$
S(t) = 0 \tag{14}
$$

So, using property 1, $D^\alpha S(t) = 0$. Therefore:

$$
D^\alpha S(t) = D^\alpha X + K(|X|^p)sign(X) = 0 \tag{15}
$$

The above equation can be rewritten in the form:

$$
D^\alpha X(t) = -K(|X|^p)sign(X) \tag{16}
$$

Based on the frequency distributed model theorem, this sliding mode dynamic is stable, and its state trajectories converge to the equilibrium $X = 0$.

As a proof, due to Theorem 2, FO sliding dynamics (Equation (16)) can be expressed as:

$$
\begin{cases} \frac{\partial \Lambda(\omega,t)}{\partial t} = -\omega^2 \Lambda(\omega, t) - K|X|^p sgn(X) \\ X(t) = \int_0^\infty u(\omega)\Lambda(\omega,t)d\omega \end{cases} \tag{17}
$$

By choosing a positive Lyapunov function in the form $V_1 = \frac{1}{2} \int_0^\infty u(\omega) \Lambda^2(\omega, t) d\omega$, one has:

$$
\begin{aligned}
D^\alpha V_1 &= \int_0^\infty u(\omega) \Lambda(\omega, t) \frac{\partial \Lambda(\omega, t)}{\partial t} d\omega \\
&= \int_0^\infty u(\omega) \Lambda(\omega, t) \left[-\omega^2 \Lambda(\omega, t) - K|X|^p sgn(X) \right] d\omega \\
&= -\int_0^\infty u(\omega) \omega^2 \Lambda^2(\omega, t) d\omega - K|X|^p sgn(X) \int_0^\infty u(\omega) \Lambda(\omega, t) d\omega \\
&= -\int_0^\infty u(\omega) \omega^2 \Lambda^2(\omega, t) d\omega - K|X|^{1+p} < 0
\end{aligned}
\tag{18}
$$

So, based on Theorem 1, the FO sliding dynamics (Equation (16)) is asymptotically stable.
Now, to guarantee the existence of the sliding motion, the following robust controller is designed:

$$
U(t) = -\left[B^{-1}AX + KB^{-1}(|X|^p)sign(X) + \gamma \, B^{-1} \, sign(S) + \lambda B^{-1} |S| \tan h(S) \right]
\tag{19}
$$

In which γ, λ are vectors of positive constants.

Theorem 3. *Consider the fractional order linear system (12). If this system is controlled by the control law (19), then the state trajectories of the system will converge to the equilibrium point.*

Proof. Choose the following Lyapunov function

$$
V_2(t) = |S|
\tag{20}
$$

Applying the D^α of $V_2(t)$ and using Lemma 1, one obtains

$$
D^\alpha V_2(t) = D^\alpha |S| = sign(S) D^\alpha S
\tag{21}
$$

Substituting s_i from (13) into (21), one has

$$
D^\alpha V_2(t) = sign(S)[D^\alpha X + K(|X|^p)sign(X)]
\tag{22}
$$

Inserting $D^\alpha X$ from (12), we have

$$
\begin{aligned}
D^\alpha V_2(t) &= sign(S)[AX + BU + K(|X|^p)sign(X)] \\
&= sign(S)[AX + B\{-[B^{-1}AX + KB^{-1}(|X|^p)sign(X) + \gamma B^{-1}sign(S) \\
&\quad + \lambda B^{-1}|S| \tan h(S)]\} + K(|X|^p)sign(X)]
\end{aligned}
\tag{23}
$$

Now some simplifications, and based on Lemma 1, we obtain
Case 1: If $S > 0$, then $sign(S) = 1$ and $0 < \tan h(S) = \varepsilon < 1$, then

$$
D^\alpha V_2(t) = -[\gamma + \lambda |S|\varepsilon] < 0
\tag{24}
$$

Case 2: If $S < 0$, then $sign(S) = -1$ and $-1 < \tan h(S) = \xi < 0$, thus

$$
D^\alpha V_2(t) = [-\gamma + \lambda |S|\xi] < 0
\tag{25}
$$

Thus, according to Theorem 1, the state trajectories of the fractional order system (11) will converge to $s_i = 0$ asymptotically. □

5. Overview of the Original SCA

The Sine Cosine algorithm (SCA) is a recently introduced stochastic heuristic scheme, which is developed based on the mathematical sine and cosine functions [33,34]. The SCA starts with initial candidate solutions and improves them through fluctuating outwards and toward the targeted global solution, using sine and cosine functions as follows:

$$x_{j,\,t+1} = \begin{cases} x_{j,\,t} + \omega \times \sin{(rand)} \times \left| rand\, P_{j,\,t} - x_{j,\,t} \right| & rand < 0.5 \\ x_{j,\,t} + \omega \times \cos{(rand)} \times \left| rand\, P_{j,\,t} - x_{j,\,t} \right| & rand \geq 0.5 \end{cases} \tag{26}$$

$$\omega = a - t\frac{a}{T_{max}} \tag{27}$$

where $x_{j,\,t}$ is the current solution at tth iteration in jth dimension, $P_{j,\,t}$ is the best solution, ω is a control parameter which decreases linearly from a constant value a to 0 by each iteration, and T_{max} is the total number of iteration.

The Hybrid SCA and HS

In spite the fact that the SCA has exhibited an efficient accuracy more often than other well-known heuristic methodologies, the native algorithm is not fitting for highly complex problems because of its poor exploration capability. To ameliorate the diversification of the standard SCA, the improvisation strategy used in HS is integrated into the SCA. In this way, a component of each search agent $X = (x_1, x_2, \ldots, x_D)$ is generated by using the SCA mechanism, with a probability of the harmony memory consideration rate ($HMCR$). While with the rate of $(1 - HMCR)$, a new component is randomly generated within the range of [LB UP], where LB and UP are the lower and upper bounds of the search agent space. Moreover, with a probability of $HMCR$ multiplying the pitch adjustment rate (PAR), the surrounding space of a search agent is searched by the coefficient distance bandwidth (bw). To guarantee a quick convergence and to guarantee the quality of the search agents, the design coefficients of the SCA-HS algorithm (PAR, bw) are dynamically updated during the evaluation procedure [35], given as:

$$PAR_t = PAR_{min} + (PAR_{max} - PAR_{min}) \times t / T_{max} \tag{28}$$

$$bw_t = bw_{max}e^{\left(\frac{\ln\left(\frac{bw_{min}}{bw_{max}}\right)}{T_{max}} \times t\right)} \tag{29}$$

where PAR_{min} and PAR_{max} are the minimum and maximum pitch adjustment, respectively. Likewise, bw_{min} and bw_{max} are the minimum and maximum bandwidths, respectively.

The computational procedure of combination of the SCA with HS is depicted in Figure 4.

	The procedure computation of the SCA-HS
I.	Generate initial population randomly
II.	**while** $t < T_{max}$
III.	Update the best solution P if there a better solution
IV.	Compute the parameters $PAR(t)$, $bw(t)$ and ω
V.	**for1** each solution
VI.	**for2** $j=1$ to D
VII.	**if1** $rand < HMCR$
VIII.	**if2** $rand < 0.5$
IX.	$x_{j,t+1}=x_{j,t} + \omega \times sin\,(rand) \times \mid rand\, P_{j,t} - x_{j,t}\mid$
X.	**else2**
XI.	$x_{j,t+1}=x_{j,t} + \omega \times cos\,(rand) \times \mid rand\, P_{j,t} - x_{j,t}\mid$
XII.	**end if2**
XIII.	**if3** $rand < PAR$
XIV.	$x_{j,t+1}=X_{j,t+1} \pm rand \times bw$
XV.	**end if3**
XVI.	**else if1**
XVII.	$x_{j,t+1}=LB_j \pm rand \times (UB_j - LB_j)$
XVIII.	**end if1**
XIX.	**end for1**
XX.	**end for2**
XXI.	$t=t+1$
XXII.	**end while**
XXIII.	Return the best solution obtained so far

Figure 4. The pseudo-code of the proposed optimization algorithm.

The evolutionary algorithms (e.g., genetic, firefly, cuckoo search, etc.) merely require information about the objective function. For online setting of the controller coefficients, a proper objective function should be defined for the guidance of its search of heuristic methodologies. In this study, the objective function of Equation (30) is adopted to adjust optimally the coefficients embedded in the FOSMC controller optimally.

$$J = \int_0^{\infty} t . e_{set-point}^2(t) + \Delta u^2(t) . dt \qquad (30)$$

where $e_{set\text{-}point}$ is the error signal and Δu is the control signal.

6. Simulation and Real-Time Results

In this section, the MG, which is provided in Figure 3, is simulated in MATLAB/Simulink (R2017a, MathWorks, MA, USA, 2017) to investigate the efficiency of the FOSMC method from a systematic perspective. To confirm the applicability of the FOSMC in the context of MGs, the experimental examinations are conducted. The FOSMC-based experimental outcomes of the test MG are compared to the well-known methodologies, such as MPC and SMC. In this application, the real-time HIL testbed is established to take into account the delays and realistic errors that are not considered in the usual off-line simulation. The schematic diagram of the HIL setup is depicted in Figure 5 and the main parts of the setup are given below [34,36].

(i)   A real-time OPAL-RT simulator is used which simulates the studied MG shown in Figure 3;
(ii) For the programming host, a PC is used as the command station to execute the MATLAB/Simulink based-code on the OPAL-RT;
(iii) A router is established to connect all the setup devices in the same sub-network. In this application, the OPAL-RT is connected to the DK60 board by Ethernet ports.

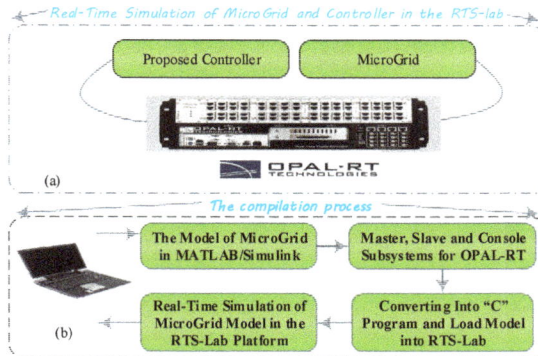

Figure 5. The real-time experimental setup; (**a**) the real-time simulation of MG and controller in the real-time simulator laboratory (RTS-LB) (**b**) the compilation process.

Case 1:

In this case, it is supposed that the load demand of the isolated MG is unvarying, i.e., $\Delta P_L = 0$. On the other hand, the power randomness of the WTG (ΔP_w) and PVG (ΔP_{pv}) are involved in the LFC-based MG. The profile of the wind power fluctuation, which is extracted from an offshore wind farm in Sweden [37], is depicted in Figure 6a, while Figure 6b illustrates the solar radiation data in Aberdeen [38], which was used in this test MG. By employing the real data, the frequency outcomes of the MG system in the HIL environment are depicted in Figure 7.

(a)

(b)

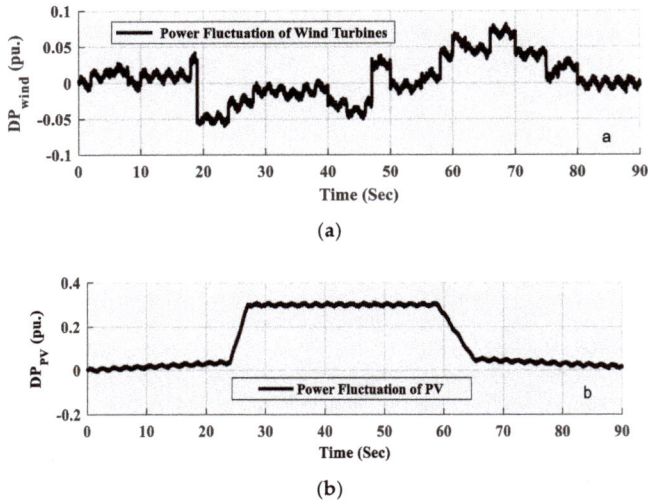

Figure 6. Power fluctuation, (**a**) wind power generator (WPG), (**b**) photovoltaic (PV).

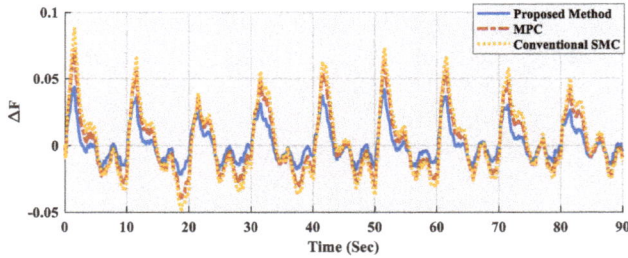

Figure 7. Curves of the MG response with the application of WPG and PV.

According to Figure 7, the desired level of the MG outcome (Δf), regarding settling time and transient fluctuations, is provided by the FOSMC controller. In comparison, the suggested technique can achieve much smaller fluctuations of Δf with quicker outcome specifications than the MPC and SMC approaches. In other words, the real-time outcomes confirm that the suggested technique can handle the randomness of the WTG and PV more effectively as compared to the considered model-based approaches.

Case 2:

In this case, to confirm the robustness of the FOSMC controller, some parameters of the MG (i.e., R, D, H, T_{inv}, and T_g) are changed. The percentage of the variations for the test system is furnished in Table 2.

As observed in Table 2, a sever scenario of changing the MG parameters is considered to assure the robustness of the suggested FOSMC scheme. The real-time results obtained by the suggested controller, MPC, and conventional SMC are depicted in Figure 8.

Table 2. Uncertain parameters of the test MG.

Parameters	Variation Range
R	−20%
D	+35%
H	−10%
T_{inv}	−25%
T_g	+25%

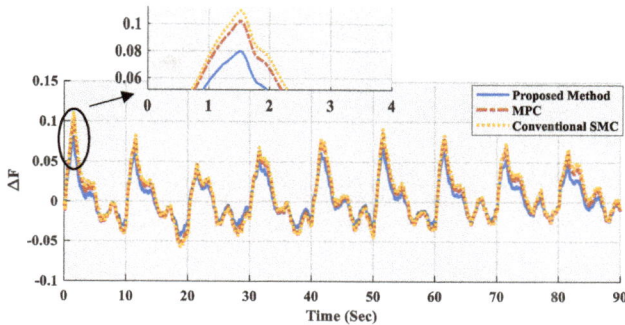

Figure 8. Curves of the MG response according to Case 2.

The results of Figure 8 reveal that when the system parameters are changed, both the MPC and conventional SMC can still provide robust LFC performance, but they are not adequately optimal. Moreover, the results indicate that the proposed controller provides a higher degree of robustness in comparison with the other two control strategies.

7. Discussion

The stochastic feature of the renewable energy items, i.e., wind and solar energies, introduces oscillations in the MG frequency. Under such circumstances, LFC plays a great role in the MG system because of its duty to preserve frequency in its scheduled value, in normal conditions, and in case of a very slight deviation of the load. However, the control of the hybrid power system operations in uncertain environments is a more complex task, without using a sufficient analytical model. This makes it necessary to apply advanced control methodologies for the realization of the system stability requirements.

Having knowledge of all the aforesaid, a new model-based FOSMC controller is developed to ameliorate the LFC performance of an MG with high penetration of RESs. To ascertain the superiority of the suggested model-based controller, two scenarios (Case 1 and Case 2) corresponding to the fluctuation of RESs and their severe parametric variations (robustness analysis) are applied. From the experimental results of Case 1 and Case 2, it is reflected that in spite of having the high system complexity with the fluctuation nature of the RESs, all the designed LFC controllers can stabilize the grid frequency effectively. In comparison, the FOSMC outperforms the MPC and conventional SMC in terms of the settling time and overshoot. The peak overshoot and undershoot of frequency deviation, using the different controllers, for the two concerned scenarios are compared, as illustrated in Figure 9. Besides, the percentage of improvement of the suggested model-based controller over the MPC and conventional SMC is depicted in Figure 10. The bar graphs of Figures 9 and 10 prove the dominance of the FOSMC controller over both of the other controllers. By comparing the results of Case 1 and Case 2, it is shown that the performance of the designed LFC controllers is deteriorated when the parametric variation is applied to the test MG. However, a higher level of stability is achieved by the proposed method than the other compared controllers.

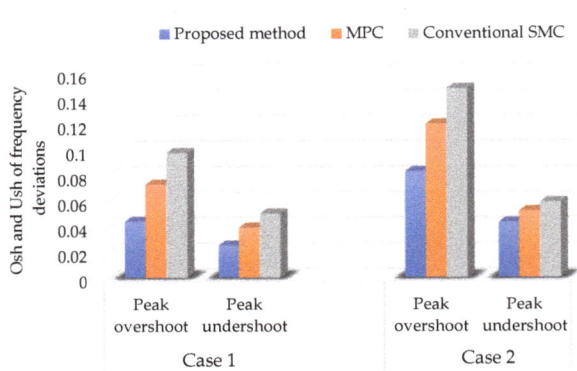

Figure 9. Peak overshoot and undershoot of frequency deviation using various control strategies.

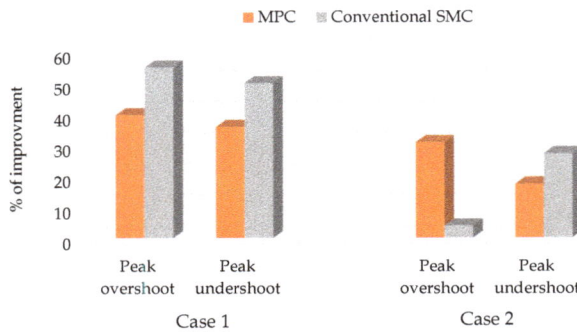

Figure 10. Percentage of improvement of the suggested technique over the MPC and conventional SMC controllers.

8. Conclusions

This work investigates a hybrid microgrid, in which the DGs and RESs are integrated to construct a complex plant with advanced functionality. A novel model-based FOSMC technique is designed for the LFC of the concerned MG with high penetration of renewable systems. Since the effectiveness of the model-based technique depends on the controller's coefficients, a hybrid SCA-HS algorithm is suggested and implemented to tune the coefficients optimally. In the sense of the Lyapunov criterion, the theoretical analysis is conducted to guarantee the stability of the suggested LFC-based MG. Furthermore, hardware-in-the-loop experiments have been carried out in this study to justify the feasibility of the FOSMC controller in a real-time environment. To show the supremacy of the suggested technique, the dynamic behavior of the test MG with the FOSMC controller is compared with the MPC and SMC approaches. Experimental outcomes confirm that the suggested technique successfully handles the aforementioned challenges of the MG, and outperforms the considered techniques.

Author Contributions: Z.E. and M.R conceived and designed the controllers; M.G wrote the paper and analyzed the data; T.D. contributed materials/analysis tools; M.-H.K. performed the experiments.

Funding: This research received no external funding.

Conflicts of Interest: The authors declare no conflict of interest.

Appl. Sci. **2019**, *9*, 3411

Abbreviations

MG	Microgrid
LFC	Load Frequency Control
SMC	Sliding Mode Control
MPC	Model Predictive Control
SCA	Sine Cosine Algorithm
HS	Harmony Search
HMCR	Harmony Memory Consideration Rate
PAR	Pitch Adjustment Rate
RES	Renewable Energy Source
DG	Distributed Generator
DEG	Diesel Engine Generator
RES	Renewable Energy Source
PV	Photovoltaic
WTG	Wind Turbine Generator
ESS	Energy Storage System
FC	Fuel Cell
FESS	Flywheel Energy Storage System
BESS	Battery Energy Storage System
PHEV	Plug-In Hybrid Electric Vehicle
MGDS	Microgrid Dispatch System
DMS	Distribution Management System
HIL	Hardware-in-the-Loop

References

1. Chowdhury, S.; Chowdhury, S.P.; Crossley, P. *Microgrids and Active Distribution Networks*; Institution of Engineering and Technology: London, UK, 2009.
2. Gheisarnejad, M.; Khooban, M.H.; Dragicevic, T. The Future 5G Network Based Secondary Load Frequency Control in Maritime Microgrids. *IEEE J. Emerg. Sel. Top. Power Electron.* **2019**. [CrossRef]
3. Gheisarnejad, M.; Khooban, M.H. Secondary load frequency control for multi-microgrids: HiL real-time simulation. *Soft Comput.* **2019**, *23*, 5785–5798. [CrossRef]
4. Bajpai, P.; Dash, V. Hybrid renewable energy systems for power generation in stand-alone applications: A review. *Renew. Sustain. Energy Rev.* **2012**, *16*, 2926–2939. [CrossRef]
5. Torreglosa, J.P.; García, P.; Fernández, L.M.; Jurado, F. Energy dispatching based on predictive controller of an off-grid wind turbine/photovoltaic/hydrogen/battery hybrid system. *Renew. Energy* **2015**, *74*, 326–336. [CrossRef]
6. Serban, I.; Teodorescu, R.; Marinescu, C. Energy storage systems impact on the short-term frequency stability of distributed autonomous microgrids, an analysis using aggregate models. *IET Renew. Power Gener.* **2013**, *7*, 531–539. [CrossRef]
7. Bevrani, H.; Feizi, M.R.; Ataee, S. Robust Frequency Control in an Islanded Microgrid: H_∞ and μ-Synthesis Approaches. *IEEE Trans. Smart Grid* **2015**, *7*, 706–717. [CrossRef]
8. Khalghani, M.R.; Khooban, M.H.; Mahboubi-Moghaddam, E.; Vafamand, N.; Goodarzi, M. A self-tuning load frequency control strategy for microgrids: Human brain emotional learning. *Int. J. Electr. Power Energy Syst.* **2016**, *75*, 311–319. [CrossRef]
9. Jiayi, H.; Chuanwen, J.; Rong, X. A review on distributed energy resources and MicroGrid. *Renew. Sustain. Energy Rev.* **2008**, *12*, 2472–2483. [CrossRef]
10. Gheisarnejad, M.; Moghadam, M.; Boudjadar, J.; Khooban, M.H. Active Power Sharing and Frequency Recovery Control in an Islanded Microgrid with Nonlinear load and Non-Dispatchable DG. *IEEE Syst. J.* **2019**, *5*, 1–11. [CrossRef]
11. Heydari, R.; Gheisarnejad, M.; Khooban, M.H.; Dragicevic, T.; Blaabjerg, F. Robust and fast voltage-source-converter (vsc) control for naval shipboard microgrids. *IEEE Trans. Power Electron.* **2019**, *34*, 8299–8303. [CrossRef]

12. Pan, I.; Das, S. Kriging Based Surrogate Modeling for Fractional Order Control of Microgrids. *IEEE Trans. Smart Grid* **2015**, *6*, 36–44. [CrossRef]

13. Pan, I.; Das, S. Fractional order fuzzy control of hybrid power system with renewable generation using chaotic PSO. *ISA Trans.* **2016**, *62*, 19–29. [CrossRef]

14. Khooban, M.H.; Niknam, T.; Shasadeghi, M.; Dragicevic, T.; Blaabjerg, F. Load frequency control in microgrids based on a stochastic noninteger controller. *IEEE Trans. Sustain. Energy* **2017**, *9*, 853–861. [CrossRef]

15. Pan, I.; Das, S. Fractional order AGC for distributed energy resources using robust optimization. *IEEE Trans. Smart Grid* **2015**, *7*, 2175–2186. [CrossRef]

16. Pahasa, J.; Ngamroo, I. Coordinated control of wind turbine blade pitch angle and PHEVs using MPCs for load frequency control of microgrid. *IEEE Syst. J.* **2014**, *10*, 97–105. [CrossRef]

17. Pandey, S.K.; Mohanty, S.R.; Kishor, N.; Catalão, J.P.S. Frequency regulation in hybrid power systems using particle swarm optimization and linear matrix inequalities based robust controller design. *Int. J. Electr. Power Energy Syst.* **2014**, *63*, 887–900. [CrossRef]

18. Bevrani, H.; Habibi, F.; Babahajyani, P.; Watanabe, M.; Mitani, Y. Intelligent frequency control in an AC microgrid: Online PSO-based fuzzy tuning approach. *IEEE Trans. Smart Grid* **2012**, *3*, 1935–1944. [CrossRef]

19. Ghafouri, A.; Milimonfared, J.; Gharehpetian, G.B. Fuzzy-adaptive frequency control of power system including microgrids, wind farms, and conventional power plants. *IEEE Syst. J.* **2014**, *12*, 2772–2781. [CrossRef]

20. Qu, Q.; Zhang, H.; Yu, R.; Liu, Y. Neural network-based H∞ sliding mode control for nonlinear systems with actuator faults and unmatched disturbances. *Neurocomputing* **2018**, *275*, 2009–2018. [CrossRef]

21. Ngo, Q.H.; Nguyen, N.P.; Nguyen, C.N.; Tran, T.H.; Ha, Q.P. Fuzzy sliding mode control of an offshore container crane. *Ocean Eng.* **2017**, *140*, 125–134. [CrossRef]

22. Ma, X.; Sun, F.; Li, H.; He, B. Neural-network-based sliding-mode control for multiple rigid-body attitude tracking with inertial information completely unknown. *Inf. Sci.* **2017**, *400*, 91–104. [CrossRef]

23. Gheisarnejad, M.; Khooban, M.H. Design an optimal fuzzy fractional proportional integral derivative controller with derivative filter for load frequency control in power systems. *Trans. Inst. Meas. Control.* **2019**, *41*, 2563–2581. [CrossRef]

24. Khooban, M.H.; Gheisarnejad, M.; Vafamand, N.; Boudjadar, J. Electric Vehicle Power Propulsion System Control Based on Time-Varying Fractional Calculus: Implementation and Experimental Results. *IEEE Trans. Intell. Veh.* **2019**, *4*, 255–264. [CrossRef]

25. Xiong, L.; Wang, J.; Mi, X.; Khan, M.W. Fractional order sliding mode based direct power control of grid-connected DFIG. *IEEE Trans. Power Syst.* **2017**, *33*, 3087–3096. [CrossRef]

26. Dadras, S.; Momeni, H.R. Fractional terminal sliding mode control design for a class of dynamical systems with uncertainty. *Commun. Nonlinear Sci. Numer. Simul.* **2012**, *17*, 367–377. [CrossRef]

27. Podlubny, I. *Fractional Differential Equations: An Introduction to Fractional Derivatives, Fractional Differential Equations, to Methods of Their Solution and Some of Their Applications*; Elsevier Science: Amsterdam, The Netherlands, 1998.

28. Li, C.; Deng, W. Remarks on fractional derivatives. *Appl. Math. Comput.* **2007**, *187*, 777–784. [CrossRef]

29. Li, Y.; Chen, Y.Q.; Podlubny, I. Stability of fractional-order nonlinear dynamic systems: Lyapunov direct method and generalized Mittag–Leffler stability. *Comput. Math. Appl.* **2010**, *59*, 1810–1821. [CrossRef]

30. Wang, B.; Ding, J.; Wu, F.; Zhu, D. Robust finite-time control of fractional-order nonlinear systems via frequency distributed model. *Nonlinear Dyn.* **2016**, *85*, 2133–2142. [CrossRef]

31. Khooban, M.-H.; Niknam, T.; Blaabjerg, F.; Davari, P.; Dragicevic, T. A robust adaptive load frequency control for micro-grids. *ISA Trans.* **2016**, *65*, 220–229. [CrossRef]

32. Khooban, M.H.; Niknam, T.; Blaabjerg, F.; Dragičević, T. A new load frequency control strategy for micro-grids with considering electrical vehicles. *Electr. Power Syst. Res.* **2017**, *143*, 585–598. [CrossRef]

33. Mirjalili, S. SCA: A Sine Cosine Algorithm for solving optimization problems. *Knowl. Based Syst.* **2016**, *96*, 120–133. [CrossRef]

34. Khooban, M.H.; Gheisarnejad, M.; Vafamand, N.; Jafari, M.; Mobayen, S.; Dragicevic, T. Robust Frequency Regulation in Mobile Microgrids: HIL Implementation. *IEEE Syst. J.* **2019**. [CrossRef]

35. Xiang, W.L.; An, M.Q.; Li, Y.Z.; He, R.C.; Zhang, J.F. An improved global-best harmony search algorithm for faster optimization. *Expert Syst. Appl.* **2014**, *41*, 5788–5803. [CrossRef]

36. Khooban, M.H.; Dragicevic, T.; Blaabjerg, F.; Delimar, M. Shipboard microgrids: A novel approach to load frequency control. *IEEE Trans. Sustain. Energy* **2017**, *9*, 843–852. [CrossRef]

37. Database of Wind Characteristics. Available online: www.winddata.com (accessed on 10 October 2014).

38. Solar Radiation Modeling. Available online: www.solargis.info/doc/solar-and-pv-data (accessed on 10 October 2014).

MDPI

St. Alban-Anlage 66

4052 Basel

Switzerland

Tel. +41 61 683 77 34

Fax +41 61 302 89 18

www.mdpi.com

Applied Sciences Editorial Office

E-mail: applsci@mdpi.com

www.mdpi.com/journal/applsci

www.ingramcontent.com/pod-product-compliance
Lightning Source LLC
Chambersburg PA
CBHW051915210326
41597CB00033B/6150